Office for Disarmament Affairs

Study on a Treaty Banning the Production of Fissile Material for Nuclear Weapons or Other Nuclear Explosive Devices

United Nations
New York, 2015

GUIDE TO THE USER

This print publication has been issued in implementation of the United Nations Disarmament Information Programme as a handy, convenient and attractive reference tool. It contains the report of the Secretary-General on the Study on a Treaty Banning the Production of Fissile Material for Nuclear Weapons or Other Nuclear Explosive Devices, as well as additional material related to the publication of the report.

The publication continues the Disarmament Study Series and serves as a valuable addition to the reference section of public and university libraries, permanent missions, research institutes and specialized non-governmental organizations.

For the electronic version of all material contained in the Study Series, see http://www.un.org/disarmament/publications/studyseries/.

Symbols of United Nations documents are composed of capital letters combined with figures. Mention of such a symbol indicates a reference to a United Nations document.

UNITED NATIONS PUBLICATION
Sales No. E.16.IX.2
ISBN: 978-92-1-142308-2
eISBN: 978-92-1-057804-2

Contents

Part 1
A/70/81
Report of the Group of Governmental Experts to make recommendations on possible aspects that could contribute to but not negotiate a treaty banning the production of fissile material for nuclear weapons or other nuclear explosive devices 1

Foreword by the Secretary-General 2

Letter of transmittal 3

I. Introduction and background overview 8

II. Treaty objectives 9

III. General characteristics and basic principles of a treaty 10

IV. Consideration of possible treaty aspects and their dynamic interrelationships 11

V. Conclusions and recommendations 24

Part 2
Background material 27

I. Resolutions. 27

A. A/RES/67/53 27

B. A/RES/66/44 29

C. A/RES/65/65 31

D. A/RES/64/29 32

E. A/RES/59/81 33

F. A/RES/58/57 34

G. A/RES/57/80 35

H. A/RES/56/24 J 36

I. A/RES/55/33 Y 37

J. A/RES/53/77 I 38

K. A/RES/48/75 L. 39

II. Report of Ambassador Shannon
CD/1299. 40

III. Views of Member States
A/68/154 and A/68/154/Add.1 42

Part 1

A/70/81

Report of the Group of Governmental Experts to make recommendations on possible aspects that could contribute to but not negotiate a treaty banning the production of fissile material for nuclear weapons or other nuclear explosive devices

Summary

The present report of the Group of Governmental Experts, established on the basis of General Assembly resolution 67/53, outlines the details of the Group's deliberations, characterizes the range of expert views on aspects of a treaty—notably in relation to the dynamic correlation between a future treaty's scope, definition, verification requirements and associated legal obligations and institutional arrangements—and presents the Group's conclusions and recommendations.

The Group reaffirmed that a treaty banning the production of fissile material for nuclear weapons or other nuclear explosive devices should be legally binding, non-discriminatory, multilateral and internationally and effectively verifiable, and that document CD/1299, and the mandate contained therein, remains the most suitable basis on which future negotiations can commence without further delay in the Conference on Disarmament and, as noted in that report, allows negotiators to raise for consideration all aspects of a treaty, including its scope. The Group agreed that such a treaty could contribute practically to achieving a world without nuclear weapons, to non-proliferation in all its aspects and, more broadly, to enhancing global security.

Experts agreed that their report, and the deliberations which underpin it, can serve as a valuable reference for States and should be a useful resource for negotiators of a future treaty. It identifies areas of convergence and divergence on key treaty aspects, including where a spectrum of views may exist and where further technical and/or scientific work can be pursued that may assist negotiators.

Foreword by the Secretary-General

The Group of Governmental Experts established by the General Assembly in its resolution 67/53, to make recommendations on possible aspects that could contribute to but not negotiate a treaty banning the production of fissile material for nuclear weapons or other nuclear explosive devices met in Geneva in four sessions of two weeks each, in 2014 and 2015.

The Group heard presentations from experts on a range of potential aspects of a future treaty and also considered the views conveyed to the Secretary-General in replies received from 17 States not members of the Group and from the European Union (see A/68/154 and Add.1). There was wide agreement that a future treaty should remain a priority of the international non-proliferation and disarmament community.

The report of the Group indicates a number of issues on which the views of most, if not all, of the experts were quite similar. There were also issues where several differing perspectives were shared and a few where positions diverged significantly.

By undertaking a fact-based and policy-neutral analysis of all aspects of a future treaty, the report of the Group constitutes an added value to the work of subsequent negotiators of a treaty.

The Group has identified the Conference on Disarmament as the venue of choice for future negotiations. Once again, I urge the Conference to adopt, without further delay, a balanced programme of work that would allow an early commencement of negotiations in light of the useful conclusions of the Group.

I take this opportunity to thank the Chair, Ambassador Elissa Golberg (Canada), and all the experts for their diligent work, which will be a useful resource for future negotiators.

Letter of transmittal

I have the honour to submit herewith the report of the Group of Governmental Experts tasked to make recommendations on possible aspects that could contribute to but not negotiate a treaty banning the production of fissile material for nuclear weapons or other nuclear explosive devices on the basis of document CD/1299 and the mandate contained therein. The Group, which the Secretary-General appointed pursuant to paragraph 3 of General Assembly resolution 67/53, comprised the following experts:

Argentina

Ms. Mariela **Fogante**
Counsellor
International Organizations Department
Ministry of Foreign Affairs and Worship

Australia

Mr. Peter **Woolcott**
Ambassador and Permanent Representative of Australia to the United Nations Office and to the Conference on Disarmament in Geneva
(First session)

Mr. John **Quinn**
Ambassador and Permanent Representative of Australia to the United Nations Office and to the Conference on Disarmament in Geneva
(Second to fourth sessions)

Brazil

Mr. João Marcelo **Galvão de Queiroz**
Head
Division of Disarmament and Sensitive Technologies
Ministry of External Relations

Canada

Ms. Elissa **Golberg**
Assistant Deputy Minister
Partnerships for Development Innovation
Foreign Affairs, Trade and Development Canada

China

Mr. Jian **Shen**
Counsellor
Permanent Mission of China to the United Nations Office and other international organizations in Geneva

Czech Republic

Mr. Michal **Merxbauer**
Director
Department of Non-proliferation
State Office for Nuclear Safety

Egypt

Mr. Hossam Eldeen **Aly**
Ambassador, Director
Disarmament Affairs
Ministry of Foreign Affairs

Finland

Mr. Olli **Heinonen**
Senior Fellow
Belfer Center for Science and International Affairs
Harvard Kennedy School, Cambridge, United States of America

France

Mr. Jean-Hugues **Simon-Michel**
Ambassador and Permanent Representative of France to the Conference on Disarmament, Geneva

Germany

Ms. Christiane **Hohmann**
Head
Division of Nuclear Arms Control, Disarmament and Non-proliferation
German Federal Foreign Office

Hungary

Ms. Judit **Körömi**
Special Representative of the Minister for Foreign Affairs and Trade for Arms Control, Disarmament and Non-Proliferation
Ministry of Foreign Affairs and Trade
(First session)

Mr. György **Molnár**
Ambassador and Special Representative of the Minister of Foreign Affairs and Trade for Arms Control, Disarmament and Non-Proliferation
Ministry of Foreign Affairs and Trade
(Second to fourth sessions)

India

Mr. Amandeep Singh **Gill**
Joint Secretary
Disarmament and International Security Affairs Division
Ministry of External Affairs

Mr. D. B. **Venkatesh Varma**
Alternate expert
Ambassador and Permanent Representative of India to the Conference on Disarmament, Geneva

Indonesia

Mr. Andy **Rachmianto**
Director
International Security and Disarmament
Ministry of Foreign Affairs

Italy

Mr. Giovanni **Manfredi**
Ambassador
Ministry of Foreign Affairs

Japan

Mr. Akio **Suda**
Ambassador and Special Assistant to the Minister for Foreign Affairs

Kazakhstan

Mr. Timur **Zhantikin**
Deputy Chair
Committee for Atomic and Energy Supervision and Control
Ministry of Energy

Mexico

Ms. Perla **Carvalho**
Ambassador and Special Adviser for Security, Disarmament and NonProliferation Issues
Office of the Vice Minister for Multilateral Affairs and Human Rights
Secretariat of Foreign Affairs
(First session)

Mr. Jorge **Lomónaco**
Ambassador and Permanent Representative of Mexico to the United Nations Office and other international organizations in Geneva
(Second session)

Mr. Rodrigo **Pintado Collet**
Senior Adviser for International Security and Disarmament Affairs
Office of the Vice Minister for Multilateral Affairs and Human Rights
Secretariat of Foreign Affairs
(Third and fourth sessions)

Netherlands

Mr. Piet **de Klerk**
Ambassador at Large
Ministry of Foreign Affairs

Nigeria

Mr. Chuka C. **Udedibia**
Ambassador and Director
Africa Multilateral Affairs Department
Ministry of Foreign Affairs

Republic of Korea

Mr. Il **Park**
Director
Disarmament and Non-proliferation Division
Ministry of Foreign Affairs
(First and second sessions)

Mr. Jong Kwon **Youn**
Director
Disarmament and Non-proliferation Division
Ministry of Foreign Affairs
(Third and fourth sessions)

Russian Federation

Mr. Mikhail I. **Ulyanov**
Director
Department for Non-proliferation and Arms Control
Ministry of Foreign Affairs

South Africa

Mr. Johann **Kellerman**
Director
Disarmament and Non-proliferation
Department of International Relations and Cooperation

Ukraine

Mr. Volodymyr **Yelchenko**
Ambassador of Ukraine to the Russian Federation

United Kingdom of Great Britain and Northern Ireland

Mr. Matthew **Rowland**
Ambassador and Permanent Representative of the United Kingdom to the Conference on Disarmament, Geneva

United States of America

Mr. Jeffrey **Eberhardt**
Director
Office of Multilateral and Nuclear Affairs
Bureau of Arms Control, Verification and Compliance
Department of State

The Group was ably assisted by staff of the United Nations Secretariat, led by Ivor Fung, and received valued expert technical advice and support from Mark Finaud and Pavel Podvig of the United Nations Institute for Disarmament Research (UNIDIR). The Group, which was unusually large to be inclusive and best account for equitable geographic representation, met in Geneva over eight-weeks during 2014 and 2015 under the auspices of the United Nations.

The Group was established to create a space where serious, substantive discussion on all possible elements of a treaty could occur, notably in the absence of negotiations having thus far commenced at the Conference on Disarmament, despite this being the overwhelming will of the States Members of the United Nations for the past 20 years. The Group succeeded in its task, conducting a robust, fact-based assessment, where no issue was off the table. The commitment to genuine dialogue apparent among its unique membership is a model for other multilateral nuclear non-proliferation and disarmament forums.

The Group's report aims to inform negotiators of a treaty, including by outlining considerable areas of convergence and on how they might address divergent perspectives. It also identifies issues where further technical/scientific work can be pursued or where ancillary confidence-building measures and/or evolutionary clauses could be developed that might assist negotiators. Ideally, the report will enjoy a wide readership and foster greater understanding of the key issues at play.

The continued value of this treaty, one that remains a logical counterpoint to the Comprehensive Nuclear-Test-Ban Treaty, has been amply demonstrated by the Group's work. We can only hope that the renewed interest and momentum generated by the work of the Group—which was evident not least in the briefings the Chair provided to the Conference on Disarmament and the General Assembly—will translate into action and the commencement of negotiations without further delay.

I have been asked by the Group to submit to you, on its behalf, the present report, which was adopted unanimously.

(*Signed*) Elissa **Golberg**
Chair of the Group

I. Introduction and background overview

1. Recognizing the essential role of fissile material in the manufacture of nuclear weapons or other nuclear explosive devices, the international community has long sought to negotiate a treaty that would ban its production for such purposes as a means to promote international nuclear disarmament and nuclear non-proliferation. In fact, the first references to the value and need for such a treaty date back more than 60 years, and since then this message has continued to be broadly and regularly reinforced. The need to ban the production of fissile material for nuclear weapons or other nuclear explosive devices was recognized, inter alia, by the General Assembly at its first special session devoted to disarmament (1978) and in numerous General Assembly resolutions, including resolution 48/75 L (1993), which called for a "nondiscriminatory, multilateral and internationally and effectively verifiable treaty". The Conference on Disarmament agreed to this mandate in the report of its Special Coordinator of 24 March 1995 (CD/1299), which was subsequently reaffirmed in the decisions of the Conference to establish subsidiary bodies to negotiate such a treaty in 1998 and 2009. In 2000, States Parties to the Treaty on the Non-Proliferation of Nuclear Weapons called for the negotiation of such a treaty at the Conference on Disarmament, "taking into consideration both nuclear disarmament and nuclear non-proliferation objectives" as part of its 13 practical steps, a call that was renewed in action 15 of the 2010 action plan on nuclear disarmament adopted by the 2010 Review Conference of the Parties to the Treaty on the Non-Proliferation of Nuclear Weapons.[1]

2. In 2012, consistent with the goal of encouraging constructive forward momentum on the issue, and acknowledging that the Conference on Disarmament had yet to commence such negotiations, the General Assembly, in its resolution 67/53, requested the Secretary-General to establish a group of governmental experts mandated to make recommendations on possible aspects that could contribute to but not negotiate a treaty banning the production of fissile material for nuclear weapons or other nuclear explosive devices,[2] on the basis of document CD/1299 and the mandate contained therein. The Group of Experts met over four two-week sessions in Geneva during 2014 and 2015. Membership was comprised of experts from 25 States chosen on the basis of equitable geographic representation.

3. In accordance with General Assembly resolution 67/53, the Group operated on the basis of consensus and during its work, reflected on the report of the Secretary-General containing the views of Member States on this subject (A/68/154). The Group took into account past consideration of such a treaty, and also requested and received informal briefings from representatives of the International Atomic Energy Agency (IAEA), the Organization for the Prohibition of Chemical Weapons and the Provisional Technical Secretariat of the Comprehensive Nuclear-Test-Ban Treaty Organization. A high degree of granularity and nuance was pursued by the Group, and emphasis was placed on the substantive legal and technical implications of different aspects of a treaty, while bearing in mind the wider context in which such an instrument would exist. The present report, which should be considered in its totality, therefore outlines the details of the Group's discussions, characterizes the range of expert views and presents the Group's conclusions and recommendations. Experts believe that their reflections can serve as useful "sign posts" for future negotiators of a treaty, without prejudice to national positions.

---

[1] See 2010 Review Conference of the Parties to the Treaty on the Non-Proliferation of Nuclear Weapons, Final Document, vols. I-III (NPT/CONF.2010/50 (Vols. I-III)); (Vol. I), Part I, sect. E.

[2] Referred here after variously as "the treaty" or "a treaty" for the sake of simplicity and without prejudice to the treaty's final form.

II. Treaty objectives

4. The Group highlighted the importance of identifying a clear object and purpose for the treaty, which would help guide negotiators in determining its scope, relevant definitions, verification requirements and associated legal obligations.

5. There was consensus that a treaty should establish a legally binding, nondiscriminatory, multilateral and internationally and effectively verifiable ban on the production of fissile material for nuclear weapons or other nuclear explosive devices. Many[3] experts suggested that a ban, along with the verification provisions of the treaty, could contribute to nuclear non-proliferation and nuclear disarmament efforts, and lay a practical foundation for additional disarmament efforts.

6. Some experts argued that in addition to a ban on new production, a treaty should also seek to prevent any increase in the amount of fissile material assigned for use in nuclear weapons or other nuclear explosive devices. These experts also felt a non-increase could be achieved by including provisions prohibiting diversion, transfers and acquisition for proscribed purposes and, separately, by ensuring the irreversibility of current and future disarmament efforts. Other experts noted the complementary role of existing legal obligations in limiting such activities and thus argued explicit prohibitions were unnecessary as these concerns would be addressed by the treaty verification regime. Some argued the concept of non-increase would only be supported in a treaty that addresses pre-existing fissile material through the establishment of a baseline to assess diversion, and for some, as a benchmark for future reductions. Some experts suggested the concept of non-increase lacked clarity and could be difficult to reflect in a treaty, or only apply to future production.

7. For some experts, the treaty should, furthermore, also address past production of fissile material as defined in the treaty itself, in order to seek to reduce and/or eliminate pre-existing fissile material accessible for additional nuclear weapons or other nuclear explosive devices. In their view, absent such measures, a treaty would neither effectively and irreversibly advance nuclear disarmament nor provide adequate incentives to join the treaty. Other experts argued that such objectives did not correspond to their understanding of the mandates contained in General Assembly resolution 48/75L and in document CD/1299, and could result in a treaty that would not achieve sufficient support or be effectively verifiable. They said that only future production of fissile material should be subject to the treaty. Some argued that past production is better addressed separately through voluntary measures and/or in the context of future disarmament negotiations.

8. When identifying the treaty's object and purpose, the Group noted that future negotiators would need to consider the interface with broader nuclear nonproliferation and nuclear disarmament efforts, which, for some experts, included consistency with existing legal obligations and instruments. Many experts stressed in this regard that a treaty should contribute to the implementation of the Treaty on the Non-Proliferation of Nuclear Weapons, including its article VI. For some experts, such a treaty should be seen as a transitional step, which would lead to eventual negotiation of a nuclear weapons convention. Others emphasized that it formed one

[3] In order to ensure equitable geographic representation and diversity of views, the General Assembly, by its resolution 67/53, established an unusually large Group, which met for an extended duration. For this reason, combined with its desire to accurately convey the degree of granularity that characterized its discussions, the Group adopted an approach of reflecting the views of its members (i.e. "most", "many" "some", "few"), which was not intended to be precedent-setting nor does it prejudice the consensus mandate that the Assembly conferred on the Group's work.

part of a framework of interlocking measures to achieve nuclear disarmament. A few experts felt that a treaty should be seen in the context of efforts to promote international stability and should be based on the principle of increased and undiminished security for all. A few experts noted that, in view of existing moratoriums introduced by some major producers of fissile material and wide adherence to the Treaty on the Non-Proliferation of Nuclear Weapons, many benefits foreseen when such a treaty was proposed had already been achieved. Some experts thought that these latter benefits referred only to non-proliferation. Some experts believed a narrower approach to developing treaty objectives should be taken, focused on non-proliferation in all its aspects, consistent with General Assembly resolution 48/75L, and stressed the importance of a stand-alone instrument that includes a verifiable ban on the production of fissile material for nuclear weapons and other nuclear explosive devices.

III. General characteristics and basic principles of a treaty

9. The Group reaffirmed that document CD/1299 and the mandate contained therein remain the most suitable basis on which future negotiations should proceed at the Conference on Disarmament and, as noted in the Shannon report, would allow negotiators to raise for consideration all aspects of a treaty, including its scope.

10. Experts agreed that an internationally and effectively verifiable treaty is one that can provide credible assurance that all States parties are complying with their treaty obligations. Moreover, a treaty should observe the principle of nondiscrimination, including through its provisions on scope, definitions and verification. Experts agreed that a treaty would be non-discriminatory if its obligations were applied equally to all States parties. Most experts recognized that the means (e.g. tools and techniques) by which these obligations are verified may vary according to the facilities located in any State party, including to account for sensitive information, but that to achieve a credible treaty such verification was necessary. Some experts believed that the non-discrimination principle should also aim to rectify perceived inequities under the Treaty on the Non-Proliferation of Nuclear Weapons with regard to safeguards obligations. Others either did not agree that such inequities exist or suggested that a treaty that sought to remedy elements present in other existing treaties and regimes would necessarily be discriminatory.

11. There was strong support for advancing the principle of irreversibility in a treaty (i.e. a one-way street), whereby steps taken to place treaty obligated fissile material and facilities that might be used to produce it under international verification cannot be reversed for the purposes of the treaty (subject to agreed termination criteria, such as irradiation of fissile material or decommissioning of a facility). Some experts also argued that irreversibility should apply to disarmament measures.

12. Many experts also saw value in the treaty's potential role in increasing transparency and confidence between States. Moreover, many experts felt that verification methods, tools and techniques developed and applied by the treaty could contribute to a broader disarmament verification methodology, as the obligations would be applied equally to all States parties, including those with currently unsafeguarded nuclear facilities.

IV. Consideration of possible treaty aspects and their dynamic interrelationships

13. Scope, definitions and verification are the main aspects of a treaty and, among them, dynamic interrelationships exist. The Group felt strongly that future negotiators would need to appreciate how a change in any one aspect affects the others. In addition to these treaty aspects, experts recognized the role that legal and institutional issues would play in effectively and efficiently achieving desired treaty objectives, including the extent to which they could contribute to the entry into force and universalization of the treaty.

A. Treaty scope

14. The Group agreed that the scope of a treaty should support its objectives and adhere to its basic principles. As described in section II above, it was clear a divergence of views existed along a spectrum on whether, and to what extent, a treaty deals with fissile material produced prior to the its entry into force.

15. Experts agreed that the treaty's scope must set out the legal obligations that States parties will be required to fulfil, and that these obligations should be formulated in terms of prohibited as opposed to permitted activities. The Group agreed that a treaty must prohibit the production of fissile material for nuclear weapons or other nuclear explosive devices. This would be the treaty's underlying obligation, applying consistently to all States parties in a non-discriminatory manner.

16. Furthermore, there was widespread agreement that the potential diversion of fissile material from non-proscribed uses, such as naval propulsion, posed a threat to the object and purpose of the treaty and the legal obligations noted in paragraph 15 above. Experts discussed different approaches to this issue, namely the inclusion of an explicit prohibition against the diversion of material, or dealing with it directly in the treaty's verification regime.

17. Many experts argued that a treaty's scope should contain additional relevant obligations. These could include an undertaking not to carry out all types of transfers of fissile material for nuclear weapons or other nuclear explosive devices. Similarly, the treaty could include a specific prohibition on the acquisition of fissile material and/or technology to produce fissile material for proscribed purposes, and from providing technical or other knowledge to assist other States in the production of material for these purposes. Some experts, however, felt such additional prohibitions on future production were unnecessary as they would be covered in an effective verification regime.

18. To assist with its deliberations on scope, the Group examined various functional categories of fissile material and the verification implications for each, which could include, inter alia, national security, commercial proprietary and resource requirements for verification. The functional categories considered may serve as a useful reference for future negotiators to determine if, and to what extent, each could be included in a treaty. A few experts expressed reservations about the typology, signalling that it was premature given the divergent views on this issue. For all functional categories of fissile material examined below, experts recalled that both the scope of the treaty and its verification regime would be linked to the definition of fissile material decided upon by the negotiators.

Fissile material produced after the entry into force of the treaty

19. Two categories of fissile material produced after entry into force were considered: material produced for civilian use and material produced for non-proscribed military use.

20. **Civilian use:** A treaty should not prohibit the production of fissile material for civilian use consistent with the obligations of State parties nor interfere in any other way with a State's right to peaceful uses of nuclear energy. Many experts agreed that the fissile material for civilian use and its production should be subject to verification under the treaty so as to deter and detect its diversion for proscribed use.

21. **Non-proscribed military use:** A treaty should not prohibit the production of fissile material for non-proscribed military purposes. This material would, however, be subject to the treaty and States parties would require credible assurance that such material is not diverted to proscribed use. Experts pointed to specific verification challenges, and the need to develop effective solutions to verify non-diversion given the sensitive nature of this material and the relevant activity itself. Some experts believed that this issue would benefit from further scientific and technical study.

Fissile material produced prior to the entry into force of the treaty

22. Experts also examined different categories of material produced prior to the entry into force of the treaty. Many experts argued that given the sizeable amounts of fissile material already possessed by some States, a ban on new production would be insufficient to achieve the objectives of the treaty, and that past production should be addressed in some manner within the treaty's scope in order to prevent diversion to proscribed use and that it should have a greater disarmament effect in practice. As is described below, these experts fell along a spectrum with regard to the extent to which past production would be addressed. Among those who supported a treaty containing provisions on past production, there was acknowledgement that its verification may need to be managed differently than future production, and most recognized that it may not be viable to insist that all elements of past production be included in the treaty. Some experts proposed that reductions in existing fissile material stockpiles be addressed through parallel arrangements, additional protocols or voluntary measures that States could pursue during or subsequent to treaty negotiations (including within a set time frame). The usefulness of such separate initiatives was, however, questioned by some experts on the basis that they may not be verifiable. Some experts noted that their understanding of General Assembly resolution 48/75 L and document CD/1299 precluded the inclusion of past production in a treaty's scope. For this reason, these experts did not feel that the detailed review held by the Group of different categories of material produced prior to entry into force was necessary. Others believed the treaty should ban only future production, but allowed that document CD/1299 left the issue of past production to negotiators. They felt that the inclusion of past production could result in a treaty that would not achieve sufficient support, nor be effectively verifiable.

23. Some experts argued that all pre-existing fissile material, with the exception of that produced for nuclear weapons, would need to be declared and be subject to some form of verification. There was considerable debate over whether States with currently unsafeguarded facilities could accurately declare past production and whether it could be verified. Some experts highlighted the practical and technical challenges with accounting post facto for historical production spanning many decades and stressed that it may not be possible for them to provide a fully accurate accounting. These experts argued that declarations of past production, when coupled with an inability to verify them, could lead to misunderstandings and the potential for

unsubstantiated allegations of non-compliance under the treaty. Other experts, however, stated that, notwithstanding these challenges, a commitment to make declarations would serve as an important trust and confidence-building measure among State parties. Some experts said that declarations and transparency on past production of fissile material would help establish a baseline to assess non-diversion and for future disarmament efforts, even if it was not possible to verify their completeness and correctness.

24. **Produced for nuclear weapons:** Without prejudice to existing nuclear disarmament obligations under other instruments, for some experts, all or part of material produced for nuclear weapons, except that in the weapons themselves, would fall within the scope of the treaty and should be covered by treaty provisions. A few experts argued that the material in the weapons themselves should be declared though not verified. Other experts stated that it was impossible to distinguish material in weapons from other material in classified form and that this, coupled with non-proliferation and security commitments, precluded effective verification and thus such material should not be subject to the treaty. A few experts suggested that all past production fell outside treaty scope.

25. **Excess to nuclear weapons requirements:** Some experts argued that, on a voluntary basis, a State party under the treaty could designate fissile material that had been produced for weapons purposes prior to entry into force as having been transferred to the civilian or non-proscribed military domain. Many experts referred to this as "excess material". Many experts argued that once such material was designated for either civilian or non-proscribed uses, a State should make such a declaration and the material would irreversibly be made subject to appropriate treaty verification. For some experts this would form the starting point of an inventory to which further material could be added. However, other experts pointed to potential verification challenges that could emerge as material declared excess can remain in sensitive form for some time. A few experts said that placing treaty obligations on excess material could serve as a disincentive for States to make such voluntary declarations and could be circumvented as such material would simply not be so designated. A few experts objected to the concept of excess material itself. Some suggested it should be omitted from the scope. Others indicated it may inadvertently confer legitimacy to continued acquisition and production of nuclear weapons, contrary to nuclear disarmament commitments undertaken by some States.

26. **Non-proscribed military use:** Taking into account the verification challenges posed by fissile material for non-proscribed military purposes outlined in paragraph 21 above, some experts argued that all fissile material produced for non-proscribed purposes prior to the entry into force of the treaty should be declared and technical or other means developed to verify non-diversion. Other experts suggested such material is beyond the scope of the treaty.

27. **Civilian use:** Many experts argued that a treaty should address the risk that pre-existing stocks of fissile material for civilian use could be diverted for proscribed purposes. Those experts considered that such material should be subject to verification under the treaty. The viability of this proposition was, however, questioned by those who said this material should not be subject to the treaty.

B. Treaty definitions

28. Experts agreed that a treaty should define, inter alia, fissile material, fissile material production and fissile material production facilities. Definitions should be practical, scientifically and technically accurate and tailored to the specific objectives of the treaty. In other words,

treaty definitions would need to be crafted in a manner that clarifies the obligations of a treaty while allowing for viable implementation and verification. The Group recalled that, in view of their dynamic interrelationship, final definitions would have an impact on treaty scope and the verification regime, with key elements of the latter requiring precise definitions to preclude variances in interpretation or in the implementation of obligations.

29. **Fissile material:** The Group considered four possible definitions for fissile material, but did not exclude the possibility of others. They considered the advantages and drawbacks of these options, the common thread among which was that highly enriched uranium and separated plutonium are the materials that should be at the heart of this definition, given their weapons applicability. Thus, the Group considered definitions based on:

(a) The IAEA safeguards concept of special fissionable material, as outlined in article XX of its statute, focusing on plutonium-239, uranium-233, uranium enriched in the isotopes 235 or 233 and mixtures containing one or more of the foregoing;

(b) The IAEA safeguards concept of unirradiated direct use material, focusing on plutonium containing less than 80 per cent Pu-238, and highly enriched uranium (containing 20 per cent or more of the isotope uranium-235 and/or uranium-233);

(c) A treaty-specific definition of weapons grade material containing 90 per cent or more of uranium-235 or uranium-233 or plutonium containing more than 95 per cent of plutonium-239;

(d) A specific isotopic composition, to be determined during negotiations based on the scope and verification requirements of the treaty.

30. Some experts favoured the definition in article XX of the IAEA statute and argued that it was already entrenched and widely understood by States through implementation of comprehensive safeguards agreements. For them, this definition would ensure treaty credibility by providing a fuller picture of a State's nuclear activities and thus allow for maximum assurance of compliance. However, other experts argued the definition was too broad; would require extensive and complex verification, including of material that could not be directly used in nuclear weapons or other nuclear explosive devices; and would generate significant verification costs. Proponents of this definition responded that not all material captured by this definition would necessarily require the same level of verification.

31. Some experts preferred the unirradiated direct use material definition on the basis that it best accounts for those materials suited for use in nuclear weapons, and was therefore effectively positioned to meet treaty objectives. They also highlighted that this definition was based on a term used in IAEA safeguards and already understood by States. For other experts, this definition was either overly or insufficiently broad in covering the nuclear fuel cycle and use of nuclear material, thereby either decreasing the effectiveness of verification or increasing the verification burden, depending on the perspective. Some proponents of this definition recognized a potential need for limited verification or transparency measures for some material not covered by the definition.

32. A few experts advocated in favour of a treaty-specific definition of weapons grade material. They argued such a definition addressed material currently applicable to nuclear weapons and thus was practical, fit for the purpose of the treaty and could ensure cost-efficient verification without undermining treaty efficacy. In their view, no State with modern nuclear arsenals would use material of inappropriate quality to produce weapons, and the non-diversion

of nuclear material in States that are not parties to the Treaty on the Non-Proliferation of Nuclear Weapons is already under IAEA verification. However, many experts believed this definition would lack credibility as it would not cover all material that has been or could be used in nuclear weapons or other nuclear explosive devices, and would leave open the possibility of evading the basic object and purpose of the treaty by producing weapons-usable material of slightly lower grade.

33. Finally, a few experts argued that a specific definition based on an isotopic composition would achieve the objective of banning the production of fissile material for nuclear weapons while avoiding unnecessary complications regarding legitimate uses of highly enriched uranium and plutonium. Agreement during negotiation would be needed on the exact isotopic concentration that proponents indicated would be close to that used in nuclear weapons and based on the scope and verification requirements of the treaty at the time of negotiation. In addition to technical factors, these experts felt the definition of fissile material should also take into account political, legal and scientific dimensions. However, many experts emphasized that this definition lacked precision, and suggested that defining a threshold close to the isotopic composition used in nuclear weapons could increase the risk that not all material usable in nuclear weapons would be covered. Such a definition might, in their view, undermine the long-accepted IAEA term for direct use material.

34. The Group agreed that each definition proposed would, to some extent, imply different types of verification tools, requiring different facilities or parts thereof to be declared, and have different implications for the intrusiveness and for cost-efficiency that States are prepared to accept with regard to a verification regime.

35. The Group explored the merits of including neptunium and americium in an eventual treaty definition of fissile material. While it was recognized that neither neptunium nor americium are currently used in nuclear weapons, some experts expressed concern that omitting these materials might create an incentive for their use in the design of new weapons. Many experts felt that the inclusion of neptunium could be considered by negotiators, or a future treaty body, should developments warrant, noting that IAEA continues to monitor this issue, including through a voluntary reporting mechanism. With regard to the inclusion of americium, some experts argued it should be given the same consideration as neptunium, while others felt it should be omitted at this time, given its limited applicability to the production of nuclear weapons.

36. Most experts agreed that tritium, as it is not by definition a fissile material, should be excluded from the treaty. A few argued it remains an important component in many nuclear weapons and should therefore be considered for inclusion.

37. Many experts recognized the need for certain treaty definitions to be sufficiently flexible to account for future scientific and technical developments and saw value in an expedited technical amendment process that would allow States parties to review and revise definitions. However, some experts did not believe that the field would evolve quickly enough to merit special procedures for rapid updates. Some experts suggested the harmonization of a treaty's definitions with those of other existing international forums be considered, noting that the definition of fissile material chosen by negotiators could have implications not only for this treaty but also existing verification procedures under IAEA and potentially elsewhere.

38. **Fissile material production:** The Group recognized that a treaty would not ban the production of fissile material per se, but rather proscribe its production for use in nuclear weapons or other nuclear explosive devices. Many experts agreed on enrichment and

reprocessing as the key production activities for the purposes of the treaty, and these would be the only activities defined as production. Some experts stated that the definition of fissile material production should be located earlier in the fuel cycle. Some experts took the view that a broader definition of fissile material production was needed encompassing other nuclear fuel cycle activities, including upstream of enrichment and reprocessing facilities, in part to achieve nondiscrimination in the treaty.

39. **Fissile material production facilities:** This definition would depend on that selected by negotiators for fissile material production. Most experts agreed that enrichment and reprocessing facilities should be at its heart, with advocates of a narrower definition of production arguing that no other facilities need be included. Those favouring a broader definition, however, argued that facilities upstream of enrichment and reprocessing facilities (although not uranium mining and milling) should be included in the definition of production facilities. Those favouring a narrower definition raised concerns about the practicality and cost-efficiency of such an approach, which, in their view, would have only marginal value in demonstrating compliance by States parties with treaty obligations. Other experts countered that, depending on the options chosen, not every facility defined as a fissile material production facility, nor all the activities occurring in it, would need to be verified with the same frequency or intensity.

40. Some experts believed it might be necessary to define a fissile material-related facility as any facility handling fissile material, including storage or processing facilities (other than enrichment or reprocessing facilities); such facilities would necessarily be declared and monitored under the verification regime. Other experts observed that verification would follow the fissile material produced and that the facilities in which it was present need not be defined.

41. Some experts advocated a definition that would distinguish between industrial-scale and laboratory-scale facilities. Others said a treaty's definitions should capture all production facilities, irrespective of scale, to avoid a verification gap (i.e. through the cumulative production of small amounts of material at multiple facilities). Future negotiators would need to explore the benefits and drawbacks of either approach, notably the benefit to an efficient verification regime of defining a *de minimis* production capacity at which production facilities would become subject to verification, versus the risk of potential clandestine production of small quantities of fissile material.

42. The Group examined issues related to the operational status of facilities, including operating closed-down, shut-down, decommissioned and dismantled facilities. Experts diverged on whether the terms relating to facility status should be incorporated into the definition of fissile material production facilities, or whether they were better addressed through a treaty's verification regime or transparency measures outside the treaty. Some experts thought that a treaty should also define fissile material diversion, transfer and acquisition. For others, since the verification regime would deter and detect such activities, the terms need not be expressly defined in the treaty.

C. Treaty verification

43. The Group reaffirmed that a treaty banning the production of fissile material for nuclear weapons or other nuclear explosive devices must, in accordance with document CD/1299, be non-discriminatory, multilateral and internationally and effectively verifiable. To experts, this meant the treaty would deter and detect noncompliance in a timely manner, provide credible assurance that States parties are complying with their treaty obligations and guard

against frivolous and/or abusive allegations of non-compliance. Declared production and, for some experts, downstream and/or upstream facilities would be the focus of verification so as to ensure that diversion of fissile material to a prohibited activity is detected and deterred. Some experts noted that verifying the correctness and completeness of State party declarations would constitute an important factor in the effectiveness of the verification regime; that the regime should provide assurance that no undeclared fissile material production is occurring, and no undeclared fissile material production facilities exist. For a few experts the verification regime should address only declared production and facilities.

44. From a technical perspective, credible assurance would be connected to the appropriate identification and consistent application of treaty verification measures. Experts noted that existing IAEA inspection goals (such as significant quantity, detection time and detection probability) may provide a useful reference point for treaty negotiators, but recognized that they would likely need to be adjusted to suit the unique purpose and context of the treaty's verification regime. Other experts noted that verification of the treaty should not imply a mechanical application of existing verification procedures from other instruments and would need to be specific to this treaty.

45. There was strong support for a non-discriminatory verification regime, under which all States parties are subject to the same obligations. Most experts recognized that the requirements of a non-discriminatory treaty could be achieved even while tailoring verification methods, tools and techniques to facilities in a State party on the basis of specific verification objectives, contexts and challenges. A few experts stated that a verification standard specific to the treaty should be agreed and that it should be applied in a non-discriminatory manner to all State parties, in particular those affected by the treaty's obligations and responsibilities. Others stressed that the IAEA verification standards should be integrated into the treaty as a means to facilitate non-discrimination. Some experts proposed an approach that applied both "light" and "heavy" touch verification (i.e. different levels of verification intrusiveness and frequency) to different fissile material production activities and facilities depending on the non-compliance risks associated with them.

46. The Group acknowledged that, in practice, a significant increase in verification requirements would occur in States currently with unsafeguarded facilities. Some experts believed the treaty's obligations would be met in non-nuclear weapon States parties to the Treaty on the Non-proliferation of Nuclear Weapons via their commitments under that treaty, by means of their comprehensive safeguards agreement. Other experts believed that, notwithstanding its voluntary nature in the context of the safeguards system related to the Treaty on the Non-Proliferation of Nuclear Weapons, non-nuclear-weapon States should also be required to adopt an additional protocol in order to provide credible assurances that no undeclared production is taking place. Still others believed that for non-nuclear-weapon States, especially those without an additional protocol, credible measures similar, but not necessarily identical, to those contained in the additional protocol should be negotiated for treaty-specific purposes. Taking into account the principle of nondiscrimination, some experts stressed that the same verification obligations should apply to all States.

47. Most experts agreed that a treaty's verification regime should seek to achieve a practical and sustainable balance between effective and resource-efficient verification acceptable to all States parties. And while the Group reinforced the need to develop a verification regime that is attentive to managing resources as cost effectively as possible, it cautioned that efficiency should not be understood to imply that effective verification activities would not be pursued

owing to cost implications. Many experts felt that avoiding unnecessary duplication of existing international verification activities, where relevant to this treaty, could assist in this regard.

48. Experts considered the benefits and drawbacks of both focused and comprehensive approaches to treaty verification.

49. A focused approach would concentrate routine verification activities at enrichment and reprocessing facilities and at those downstream facilities that are processing or handling fissile material. Those favouring this approach spoke to its simplicity and resource efficiency. They argued that these facilities were most directly implicated in fissile material production. The verification of these facilities, coupled with monitoring of the use of fissile material produced, and supplemented by measures to detect possible undeclared production, would ensure against diversion for proscribed purposes. However, some experts considered this approach too narrow to provide for confidence of coverage under the treaty.

50. Under a comprehensive verification approach, in addition to that which is covered by the focused approach, verification activities would also cover most if not all of the nuclear fuel cycle. This would include upstream processing, which some experts pointed out would extend to other, less sensitive nuclear material. Power reactors and spent fuel would also be captured. Some experts felt that this approach would better meet treaty requirements, while others argued that the limited risk of diversion of this material did not merit the more complex and costly verification required.

51. Given the diversity in views, some experts also considered that negotiators could adopt a hybrid approach, concentrating particularly on those areas of the nuclear fuel cycle where the degree of attractiveness for misuse is highest, should a party seek to violate its treaty obligations.

52. Some experts highlighted that the level of verification, if any, that should be conducted at a given fissile material production facility would also depend on its status, of which operating, shut-down, closed-down, decommissioned or dismantled were possible categories.

53. The Group recognized that a treaty's verification regime will have to take into account the concerns of States parties regarding sensitive information, whether related to national security, non-proliferation or commercial proprietary reasons, in a manner that avoids compromising the credibility and efficacy of verification efforts. Many experts highlighted the need for greater clarity in order for all States parties to better understand where the limits to verification may stand, and emphasized that it was incumbent on an inspected State to nonetheless provide credible assurances to the international community by making best efforts to accommodate inspection requests and, if access could not be provided without compromising sensitive information, to seek to resolve questions by other means. Some experts noted this may prove difficult in certain instances, given the classification of relevant information. Some experts agreed that effective solutions would need to be developed, and suggested that useful models in this regard included a "black box" approach, focusing on limited verification of input and output and managed access procedures. Some experts felt that future negotiators could potentially benefit from further technical and scientific consideration of these issues.

54. Experts also discussed the implications of verifying material for non-proscribed military use (such as naval fuel) and associated facilities. Many experts felt that future production for non-proscribed military purposes should be verified in an appropriate manner, with some also arguing that past production declared for this purpose should be verified so as to preclude that it is used for proscribed purposes. As described in paragraph 53 above, useful

models mentioned included a "black box" approach and, for some experts, procedures based on article 14 of the IAEA model comprehensive safeguards agreement and/or managed access.

55. Similarly, some experts suggested that a treaty would require the development of verification tools to address the sensitivity of verifying production facilities that existed in States with unsafeguarded facilities prior to entry into force, which are then designated for production for non-proscribed purposes. Others argued however, that the process of converting a facility from military to non-proscribed use does not require specific verification tools since both the facility, and the material it contains, would become subject to verification.

56. Despite the objections raised in paragraph 25 above regarding the concept of material in excess to nuclear weapons requirements, experts debated potential verification approaches to address it. Many experts agreed that such material would, once in non-sensitive form, become subject to the treaty in a manner equivalent to material produced for non-proscribed purposes. Some experts argued that verification should begin (and declarations be provided) at the point when this material is declared excess, recognizing that special verification measures might be required given its sensitive nature. Other experts countered with regard to the complexities of verifying material at this early stage, and argued that verification could only begin once material has been converted to a non-sensitive form, composition and mass. Some experts also held the view that past production should remain outside the scope of a treaty that prohibits the production and not the possession of fissile material for use in nuclear weapons or other nuclear explosive devices.

57. Some experts argued that all past production of fissile material should be accounted for and verified under a treaty in order to provide a clear baseline on which to assess and verify treaty compliance. Many experts identified challenges to such an approach, with some arguing that verification of all past production may not be possible given the inability to determine a satisfactory historical accounting of the material originally produced. The Group discussed concerns about the dearth of efficient and effective technical means to conduct verification in States that are presently with unsafeguarded facilities, notably given proliferation and national security concerns, obligations under the Treaty on the Non-Proliferation of Nuclear Weapons prohibiting the transfer of sensitive information and the international community's lack of verification experience in military fuel cycle facilities. Also noted was the challenge of effectively and efficiently verifying large numbers of facilities or large-scale facilities, particularly those that are decades old and not designed to support monitoring. However, many experts argued that techniques could be developed to address these challenges.

58. **Declarations:** Experts agreed that to facilitate effective verification, States parties would be required to provide an initial declaration of fissile material production facilities and, for some experts, related downstream facilities handling fissile material and/or upstream facilities. They acknowledged, however, that the nature and content of such declarations would depend on the scope and definitions ultimately decided by negotiators. Some argued that all existing fissile material, including that produced prior to entry into force of the treaty, should be included, whereas others believed this would fall outside treaty scope. Experts discussed possible specific content for such declarations, including design information and status of facilities that produce fissile material, those that process and handle fissile material and, if applicable, the inventories of fissile material contained therein. Initial declarations would be complemented by ongoing declaration requirements to capture the production of fissile material, plans to construct new facilities or changes the status of existing ones. Some experts also noted the importance of providing design information and future development plans in order to facilitate effective verification efforts.

59. **Verification toolbox:** The Group agreed that a diverse verification toolbox would be necessary for the regime to provide credible assurance that States parties complied with their obligations and treaty verification requirements. Such a toolbox should include verification tools and techniques capable of supporting the verification standard agreed by States parties with respect to the implementation of their obligations. Most experts agreed that existing verification methods, tools and techniques employed in relevant multilateral or bilateral forums, most notably IAEA, but also, potentially, elements of those used by the Organization for the Prohibition of Chemical Weapons and the Provisional Technical Secretariat of the Comprehensive Nuclear-Test-Ban Treaty Organization, should likely form the core of the verification toolbox. Many experts felt that further study is needed in order to understand how they could apply in this treaty context, with negotiators considering the unique circumstances in which they were developed and implemented. The existing verification methods used by IAEA, including a system of material accountancy and control, were seen by most experts as being highly relevant to a treaty's toolbox, notably as a means to verify the correctness and completeness of State declarations of fissile material production and facilities. Other experts felt that existing IAEA approaches to verification were inappropriate for the treaty and that the verification toolbox could be discussed in depth only after the scope of the treaty had been determined. Experts agreed that it is important to leave flexibility in the toolbox to account for future developments in verification technology and, for some experts, the requirements of ongoing and future disarmament efforts. For a few experts, any change in the verification toolbox would need to be approved by all States parties. Given the complexity and interrelated nature of these issues, some experts noted the importance of continued consideration of verification methods, tools and techniques to add value to future negotiations.

60. Many experts highlighted the need for tools to detect undeclared production and facilities in States that currently have unsafeguarded facilities, and reflected on those presently employed by IAEA, although a few experts did not support the application of this concept to the treaty. A few experts believed that these States should adopt a comprehensive safeguards agreement and an additional protocol. Others highlighted that tools and techniques, similar to those contained in the model additional protocol could be in place in order to provide credible assurances that no undeclared production is occurring in those States. Still, a few experts stressed that a verification regime fit for purpose should be negotiated for the treaty. Environmental sampling was also suggested by some as a relevant IAEA tool when applied in the appropriate context and locations, although some experts questioned its value in detecting undeclared activity in States that had been operating unsafeguarded facilities on a significant scale, not least due to false alarms potentially generated by past production. Some experts also noted that verification challenges could be met through other measures without risking the release of sensitive information. These issues are likely to be site specific and will likely be resolved on a case-by-case basis.

61. Most experts agreed that a treaty's verification toolbox should include provisions for non-routine inspections, including challenge inspections, as a means to detect and deter undeclared fissile material production. Experts noted the value in examining various approaches to a model for non-routine inspections in the treaty, including, inter alia, IAEA special inspections and complementary access provisions, the challenge inspections procedures employed by the secretariat of the Convention on the Prohibition of the Development, Production, Stockpiling and Use of Chemical Weapons and on Their Destruction and the provisions used by the Provisional Technical Secretariat of the Comprehensive Nuclear-Test-Ban Treaty Organization for on-site inspections. Experts noted that specific procedures would be needed to launch and carry out such non-routine inspections.

D. Legal and institutional arrangements

62. The legal and institutional arrangements of a treaty banning the production of fissile material for nuclear weapons or other nuclear explosive devices will play an important role in ensuring its credibility within the international community. Experts agreed that these arrangements should be designed to facilitate the treaty's effective implementation so that it may achieve its intended object and purpose.

63. The Group counselled that negotiators would therefore need to balance credibility and flexibility while keeping in mind the need for precise and practical approaches. It was widely noted that the legal and institutional arrangements of a treaty should take into account the current and evolving experience of other international treaties and institutions. To maintain the confidence of States parties, the Group felt the treaty's institutions would need to remain credible, including by being politically impartial and technically able and by applying resource-efficient techniques. Some experts recommended that future negotiators consider ways to promote adherence to the treaty in the design of legal and institutional arrangements by increasing trust and confidence in the treaty, including through incentives.

64. **Governance:** Experts agreed that a treaty should include the establishment of governance and decision-making mechanisms that would provide political oversight and conduct treaty-related decision-making, including on issues of resource allocation, non-compliance and verification. Such mechanisms should include a conference of States parties, an executive council of some form and a secretariat. Citing membership and mandate concerns, some experts advocated an executive council, independent and unique to the treaty. Others felt that membership and mandate concerns could be addressed by establishing a separate treaty executive council within IAEA, while a few experts argued that the existing IAEA Board of Governors could fulfil this role, recognizing that it had assumed new responsibilities in the past. Many experts felt that additional consideration of the structure of these proposed mechanisms is desirable.

65. Some experts outlined the need for a treaty organization or secretariat, which would, at a minimum, provide support to the governance and decision-making bodies and manage administrative matters. It could also be tasked to monitor implementation of the treaty more broadly. There were differing perspectives on whether this entity should be independent and unique to the treaty or contained within IAEA. Some experts argued that details relating to the functioning of the treaty organization/secretariat should not be elaborated in the treaty itself, but would be best determined by the States parties at subsequent meetings.

66. Some experts proposed that the conference of State parties meet annually and that it have decision-making authority on implementation of the treaty, as well as oversight of the executive council and secretariat. Some felt this should also be complemented by a regular cycle of review conferences while others felt that periodic meetings of the conference of States parties would suffice.

67. For experts, an international and effectively verifiable treaty would be one that involves credible multilateral verification. Negotiators would need to determine whether verification responsibilities should be carried out by IAEA or be self-contained within the treaty organization/secretariat. Experts in favour of locating the verification body within IAEA suggested it would avoid unnecessary duplication and ensure the consistent application of treaty obligations for the many States already under IAEA verification mechanisms, and could be accomplished through a formal cooperation mechanism. A structure similar to the Iraq Action

Team within IAEA was noted as a possibility. Resource implications and the depth of experience and established technical capabilities of IAEA were also cited as factors in favour of its selection as a verification body. Other experts, however, felt that the distinct purpose of the treaty, and potential for different membership from IAEA, argued in favour of a self-contained verification body situated within the treaty organization/secretariat, with legal authority and resources capable of verification and other implementation functions. They noted the limited IAEA experience with verification in States currently with unsafeguarded facilities, and the potential for proliferation of sensitive information (which other experts noted would exist irrespective of the body selected). For some experts, however, even if an independent verification body were pursued, it should be positioned to draw on the useful expertise and resources of IAEA. Experts noted that any role for IAEA would have implications for its current functioning.

68. Drawing on lessons from existing arrangements in similar treaties, negotiators will need to consider how the structure and functioning of the verification regime should best be reflected in the text of a treaty. For example, some experts suggested that elaborate technical details could be integrated in parallel annexes or protocols (in a model similar to the Convention on the Prohibition of the Development, Production, Stockpiling and Use of Chemical Weapons and on Their Destruction or the Comprehensive Nuclear-Test-Ban Treaty) or in subsidiary arrangements outlined separately between the verification body and individual States parties, as in the case of IAEA safeguards. Caution was noted by some about being overly prescriptive, given the pace of technological change. Others highlighted that provisions on the protection of confidential information would be pertinent to treaty implementation.

69. **Compliance:** Experts agreed that States parties, whether through a conference of States parties or an executive council, should play the main role in assessing cases of non-compliance. For some experts both States parties and the verification body could have a role in invoking certain non-routine inspection rights, while for others such a role was limited to States parties. Experts noted the importance of establishing mechanisms within a treaty to reduce the probability and verify the credibility of non-compliance accusations that might be frivolous and/or abusive.

70. Experts agreed that a process would be needed to address allegations of noncompliance. Many experts saw value in using a cooperative approach in response to initial reports of non-compliance, which could serve as an incentive by ensuring the participation of involved parties. Some experts suggested negotiators might find value in examining existing models that use a cooperative approach, such as the Convention on the Prohibition of the Use, Stockpiling, Production and Transfer of Anti-Personnel Mines and on Their Destruction, the Convention on Prohibitions or Restrictions on the Use of Certain Conventional Weapons Which May be Deemed to be Excessively Injurious or to Have Indiscriminate Effects and the Comprehensive Nuclear-Test-Ban Treaty (consultation and clarification). Only where serious concerns are identified, and where cooperative deliberations are unproductive or inconclusive, would a finding of non-compliance be made. Some experts suggested assessments of non-compliance could be addressed within the conference of States parties. The executive council, an ad hoc committee or the conference on States parties, may have a role in each or any of these approaches.

71. The Group explored the benefits and drawbacks of having formal findings of non-compliance referred to the Security Council or the General Assembly. Experts discussed the institutional problems that could arise in a situation involving noncompliance by a permanent member of the Security Council. A few experts suggested that negotiators may wish to examine how the issue was considered in the development of the Comprehensive Nuclear-Test-Ban

Treaty. Some experts also proposed that the treaty should contain its own punitive measures (inter alia, suspension of membership from decision-making bodies), which could be imposed, as per procedures agreed in the treaty, prior and without prejudice to the penalties imposed by another principal organ of the United Nations. Finally, experts raised the potential of complications in dealing with non-compliance in States with existing IAEA obligations under multiple regimes, with possibly different executive bodies of divergent membership. Experts felt that these issues merited further consideration.

72. **Amendment and review:** A treaty should include practical amendment provisions to be pursued in conformity with international treaty law. Experts discussed different mechanisms for making amendments, which could include a referral mechanism to the conference of States parties, a review conference or an ad hoc amendment conference (called by two-thirds of the members of the conference of States parties). Some experts argued that given the potential ramifications of substantive treaty amendments, they should be adopted by consensus. Others argued that while consensus should be the goal, a treaty should adopt the practice of the most recent disarmament conventions and allow voting. The possibility of qualified voting was discussed.

73. Separately, many experts noted that depending on the detail contained in the treaty on issues such as verification, a dedicated and expedited process to address technical updates might be needed. This mechanism should be flexible and non-resource-intensive, with some experts proposing this task could fall to a review conference.

74. **Entry into force:** The Group felt that negotiators would need to consider the range of options between overly restrictive entry into force provisions that could block its effective implementation and more permissive provisions that could undermine treaty credibility. Many experts felt that proposing a simple unqualified number of States necessary for ratification would affect the treaty's credibility and that entry into force should require ratification by a specified number of States with unsafeguarded facilities. Some experts pointed to the need to set out such provisions without conferring any unintended legal status. Other experts argued that legal status would be unique to the treaty and that it should not target any country or rely on categories of States contained in other treaties. Some experts argued that ratification by a certain number, or all, States with enrichment and reprocessing capabilities should be the criteria used for entry into force. Although the number of States falling into this category would be directly impacted by the treaty definition of fissile material production, some experts noted that it could be technically challenging to determine which States would fit in this category. Another approach would be to require a specific number of States from more than one of the categories noted above. Experts agreed that treaty universality would remain an important goal.

75. **Duration:** Many experts argued that indefinite duration of the treaty would preserve the irreversibility of measures taken under it. Some experts suggested that a treaty of long duration (such as 25 years), perhaps with renewal provisions, could achieve the same objectives. While many experts were confident that either indefinite or lengthy treaty duration would sustain disarmament momentum, some viewed the treaty as an interim step and expressed concern that a treaty with indefinite duration would reduce incentives and pressure to make progress on other disarmament commitments.

76. **Withdrawal:** Consistent with international law and notwithstanding efforts to ensure irreversibility, the Group agreed that parties to a treaty should have the right to withdraw, although given its potential strategic implications and the desire to prevent potential abuse, experts felt that conditions for withdrawal should be restrictive. For example, the treaty

should require adequate advance notice for withdrawal, and members of the Group felt that States should remain responsible for commitments previously entered into on treaty obligated material, and for any violations that may have been committed prior to withdrawal. Some experts suggested that withdrawal by a State with currently unsafeguarded fissile material could risk undermining the object and purpose of a treaty, with a few indicating it should trigger its termination while others countered that such singling out was discriminatory.

77. **Other issues:** Experts recognized that other legal and institutional issues such as accession, reservations, depository, dispute settlement, etc. would benefit from further analysis. Some experts felt that legal and institutional issues are interrelated and should be addressed in an integrated manner.

V. Conclusions and recommendations

78. Given the dynamic and challenging contemporary global security environment, members of the Group believed firmly in the importance of ensuring continued international commitment and high-level attention to making practical progress on achieving a world without nuclear weapons and on non-proliferation in all its aspects. In this regard, the Group agreed that a treaty could contribute practically to this goal, and more broadly to enhancing global security.

79. The work of the Group represented the most thorough expert intergovernmental assessment to date of a treaty banning the production of fissile material for nuclear weapons or other nuclear explosive devices. Given the depth of its exchanges, the Group has allowed for a better understanding of the potential architecture of a future treaty and has further clarified its various aspects. It also identified areas, including technical and scientific challenges to definitions, verification and scope, or on legal and institutional matters, which will have implications for future treaty negotiations. In this regard, it helped identify areas of potential convergence and divergence and where a spectrum of views may exist. The Group concluded its work with a fuller appreciation of the range of expert positions, including on issues that may pose challenges to negotiators. Without prejudice to national positions, the members of the Group are confident that this report and the deliberations which underpin it, can serve as a valuable reference for States and should be a useful resource for negotiators of a future treaty.

80. In addition, pursuant to its discussions, the Group recommends that:

- Future negotiators of a treaty take into account the work of the Group, as appropriate in their deliberations.
- In conveying the work of the Group to the Conference on Disarmament, the Secretary-General call upon the Conference to consider and fully examine the report of the Group. States members of the Conference on Disarmament are encouraged to include in their delegations technical experts, as may be required, to facilitate deliberations on issues identified in the report.
- The Secretary-General, building on the 2013 report on the treaty banning the production of fissile material for nuclear weapons or other nuclear explosive devices (A/68/154), seek the views of the States Members of the United Nations on the present report.
- States give due consideration to the report of the Group and the report be made available to the wider international community and civil society, for example on the websites of the United Nations and the Conference on Disarmament.

81. Finally, the Group has demonstrated through its significant analysis and thoughtful dialogue that the various perspectives of States on a treaty should not be an obstacle to commencement of negotiations. In accordance with General Assembly resolution 67/53, the Group considers that document CD/1299 and the mandate contained therein continues to provide the most suitable basis on which future negotiations can commence without further delay in the Conference on Disarmament and, as noted in the Shannon report, would allow negotiators to raise for consideration all aspects of a treaty, including its scope. In this regard, members of the Group believe that continued active support and leadership by the Secretary-General is important to maintain momentum.

Part 2
Background material

I. Resolutions

A. A/RES/67/53
Treaty banning the production of fissile material for nuclear weapons or other nuclear explosive devices

The General Assembly,

Recalling its resolutions 48/75 L of 16 December 1993, 53/77 I of 4 December 1998, 55/33 Y of 20 November 2000, 56/24 J of 29 November 2001, 57/80 of 22 November 2002, 58/57 of 8 December 2003, 59/81 of 3 December 2004, 64/29 of 2 December 2009, 65/65 of 8 December 2010 and 66/44 of 2 December 2011 on the subject of banning the production of fissile material for nuclear weapons or other nuclear explosive devices,

Recalling also document CD/1299 of 24 March 1995, in which all members of the Conference on Disarmament agreed on the mandate to negotiate a treaty banning the production of fissile material for nuclear weapons or other nuclear explosive devices and which would not preclude any delegation from raising for consideration, in negotiations, any issue noted therein,

Mindful of the continuing importance and relevance of the Conference on Disarmament, and recalling the past achievements of that body in successfully negotiating arms control and disarmament agreements,

Recalling the support for the Conference on Disarmament expressed by the Security Council summit on nuclear disarmament and nuclear non-proliferation, held on 24 September 2009, as well as similar messages expressed at the high-level meeting on revitalizing the work of the Conference on Disarmament and taking forward multilateral disarmament negotiations, held in New York on 24 September 2010, and the follow-up plenary meeting of the General Assembly, held from 27 to 29 July 2011,

Expressing frustration with the years of stalemate in the Conference on Disarmament, and looking forward to the Conference again fulfilling its mandate as the world's single multilateral disarmament negotiating forum,

Convinced that a non-discriminatory, multilateral and internationally and effectively verifiable treaty banning the production of fissile material for nuclear weapons or other nuclear explosive devices would be a significant contribution to nuclear disarmament and non-proliferation,

Recognizing the importance of advancing all issues identified in decision CD/1864, adopted by consensus by the Conference on Disarmament on 29 May 2009,

Welcoming the discussions involving scientific experts on various technical aspects of a treaty banning the production of fissile material for nuclear weapons or other nuclear explosive devices that have taken place in Geneva, both within and on the margins of the Conference on Disarmament, in support of the early commencement of negotiations, including those meetings that took place in 2012 pursuant to resolution 66/44,

Noting the determination expressed by China, France, the Russian Federation, the United Kingdom of Great Britain and Northern Ireland and the United States of America at the meeting held in Washington, D.C., from 27 to 29 June 2012 to renew their efforts, with relevant parties, to achieve a treaty banning the production of fissile materials for nuclear weapons or other nuclear explosive devices at the earliest possible date in the Conference on Disarmament,

Noting also that the Conference on Disarmament did not succeed in adopting a programme of work by the end of its 2012 session,

1. *Urges* the Conference on Disarmament to agree on and implement early in 2013 a balanced and comprehensive programme of work that includes the immediate commencement of negotiations on a treaty banning the production of fissile material for nuclear weapons or other nuclear explosive devices on the basis of document CD/1299 and the mandate contained therein;

2. *Requests* the Secretary-General to seek the views of Member States on a treaty banning the production of fissile material for nuclear weapons or other nuclear explosive devices, including possible aspects thereof, and to submit a report on the subject to the General Assembly at its sixty-eighth session;

3. *Also requests* the Secretary-General to establish a group of governmental experts with a membership of 25 States chosen on the basis of equitable geographical representation, which, taking into account the report containing the views of Member States, will make recommendations on possible aspects that could contribute to but not negotiate a treaty banning the production of fissile material for nuclear weapons or other nuclear explosive devices on the basis of document CD/1299 and the mandate contained therein, which will operate on the basis of consensus, without prejudice to national positions in future negotiations and which will meet in Geneva for two sessions of two weeks in 2014 and in 2015;

4. *Calls upon* the Secretary-General to transmit the report of the group of governmental experts to the General Assembly at its seventieth session and to the Conference on Disarmament;

5. *Invites* the Conference on Disarmament to take note of the report of the group of governmental experts and consider further action as appropriate;

6. *Decides* that, should the Conference on Disarmament agree upon and implement a balanced and comprehensive programme of work that includes negotiation of a treaty banning the production of fissile material for nuclear weapons or other nuclear explosive devices, the group of governmental experts shall conclude and its work shall be submitted to the Secretary-General for onward transmission to the Conference on Disarmament;

7. *Also decides* to include in the provisional agenda of its sixty-eighth session, under the item entitled "General and complete disarmament", the sub-item entitled "Treaty banning the production of fissile material for nuclear weapons or other nuclear explosive devices".

48th plenary meeting
3 December 2012

B. A/RES/66/44
Treaty banning the production of fissile material for nuclear weapons or other nuclear explosive devices

The General Assembly,

Recalling its resolutions 48/75 L of 16 December 1993, 53/77 I of 4 December 1998, 55/33 Y of 20 November 2000, 56/24 J of 29 November 2001, 57/80 of 22 November 2002, 58/57 of 8 December 2003, 59/81 of 3 December 2004, 64/29 of 2 December 2009 and 65/65 of 8 December 2010 on the subject of banning the production of fissile material for nuclear weapons or other nuclear explosive devices,

Recalling also document CD/1299 of 24 March 1995, in which all members of the Conference on Disarmament agreed on the mandate to negotiate a treaty banning the production of fissile material for nuclear weapons or other nuclear explosive devices and which would not preclude any delegation from raising for consideration, in negotiations, any issue noted therein,

Recalling further the support for the Conference on Disarmament expressed by the Security Council summit on nuclear disarmament and nuclear non-proliferation, held on 24 September 2009,

Convinced that a non-discriminatory, multilateral and internationally and effectively verifiable treaty banning the production of fissile material for nuclear weapons or other nuclear explosive devices would be a significant contribution to nuclear disarmament and non-proliferation,

Recognizing the importance of advancing all issues identified in decision CD/1864, adopted by consensus by the Conference on Disarmament on 29 May 2009,

Noting the determination of China, France, the Russian Federation, the United Kingdom of Great Britain and Northern Ireland and the United States of America at the meeting held in Paris on 30 June and 1 July 2011 to renew their efforts, with relevant parties, to achieve a treaty banning the production of fissile materials for nuclear weapons and other nuclear explosive devices at the earliest possible date in the Conference on Disarmament,

Expressing frustration with the years of stalemate in the Conference on Disarmament, which has prevented it from fulfilling its mandate as the world's single multilateral disarmament negotiating forum,

1. *Urges* the Conference on Disarmament to agree on and implement early in 2012 a comprehensive programme of work that includes the immediate commencement of negotiations on a treaty banning the production of fissile material for nuclear weapons or other nuclear explosive devices on the basis of document CD/1299 and the mandate contained therein;

2. *Resolves* to consider options for the negotiation of a treaty banning the production of fissile material for nuclear weapons or other nuclear explosive devices at its sixty-seventh session should the Conference on Disarmament fail to agree on and implement a comprehensive programme of work by the end of its 2012 session;

3. *Encourages* interested Member States, without prejudice to their national positions during future negotiations on such a treaty, to continue efforts, including within and on the margins of the Conference on Disarmament, in support of the commencement of negotiations,

including through meetings involving scientific experts on various technical aspects of the treaty, drawing on available expertise from the International Atomic Energy Agency and other relevant bodies, as appropriate;

4. *Decides* to include in the provisional agenda of its sixty-seventh session the item entitled "Treaty banning the production of fissile material for nuclear weapons or other nuclear explosive devices".

71st plenary meeting
2 December 2011

C. A/RES/65/65
Treaty banning the production of fissile material for nuclear weapons or other nuclear explosive devices

The General Assembly,

Recalling its resolutions 48/75 L of 16 December 1993, 53/77 I of 4 December 1998, 55/33 Y of 20 November 2000, 56/24 J of 29 November 2001, 57/80 of 22 November 2002, 58/57 of 8 December 2003, 59/81 of 3 December 2004 and 64/29 of 2 December 2009 on the subject of banning the production of fissile material for nuclear weapons or other nuclear explosive devices, and noting in this regard the support for the Conference on Disarmament expressed by the Security Council summit on nuclear disarmament and nuclear non-proliferation, held on 24 September 2009,

Convinced that a non-discriminatory, multilateral and internationally and effectively verifiable treaty banning the production of fissile material for nuclear weapons or other nuclear explosive devices would be a significant contribution to nuclear disarmament and non-proliferation,

Welcoming, after years of stalemate, the consensus adoption by the Conference on Disarmament of its decision (CD/1864) of 29 May 2009 on the establishment of a programme of work for its 2009 session, by which the Conference, inter alia, and without prejudice to any past, present or future position, established a Working Group to negotiate a treaty banning the production of fissile material for nuclear weapons or other nuclear explosive devices on the basis of document CD/1299 of 24 March 1995 and the mandate contained therein,

1. *Urges* the Conference on Disarmament to agree early in 2011 on a programme of work that includes the immediate commencement of negotiations on a treaty banning the production of fissile material for nuclear weapons or other nuclear explosive devices;

2. *Decides* to include in the provisional agenda of its sixty-sixth session the item entitled "Treaty banning the production of fissile material for nuclear weapons or other nuclear explosive devices".

60th plenary meeting
8 December 2010

D. A/RES/64/29
Treaty banning the production of fissile material for nuclear weapons or other nuclear explosive devices

The General Assembly,

Recalling its resolutions 48/75 L of 16 December 1993, 53/77 I of 4 December 1998, 55/33 Y of 20 November 2000, 56/24 J of 29 November 2001, 57/80 of 22 November 2002, 58/57 of 8 December 2003 and 59/81 of 3 December 2004 on the subject of banning the production of fissile material for nuclear weapons or other nuclear explosive devices, and noting in this regard the support for the Conference on Disarmament expressed by the Security Council summit on nuclear disarmament and nuclear non-proliferation, held on 24 September 2009,

Convinced that a non-discriminatory, multilateral and internationally and effectively verifiable treaty banning the production of fissile material for nuclear weapons or other nuclear explosive devices would be a significant contribution to nuclear disarmament and non-proliferation,

Welcoming, after years of stalemate, the consensus adoption by the Conference on Disarmament of its decision (CD/1864) of 29 May 2009 on the establishment of a programme of work for its 2009 session, by which the Conference, inter alia, and without prejudice to any past, present or future position, established a Working Group to negotiate a treaty banning the production of fissile material for nuclear weapons or other nuclear explosive devices on the basis of document CD/1299 of 24 March 1995 and the mandate contained therein,

1. *Urges* the Conference on Disarmament to agree early in 2010 on a programme of work that includes the immediate commencement of negotiations on a treaty banning the production of fissile material for nuclear weapons or other nuclear explosive devices;

2. *Decides* to include in the provisional agenda of its sixty-fifth session an item entitled "Treaty banning the production of fissile material for nuclear weapons or other nuclear explosive devices".

55th plenary meeting
2 December 2009

E. A/RES/59/81 The Conference on Disarmament decision (CD/1547) of 11 August 1998 to establish, under item 1 of its agenda entitled "Cessation of the nuclear arms race and nuclear disarmament", an ad hoc committee to negotiate, on the basis of the report of the Special Coordinator (CD/1299) and the mandate contained therein, a non-discriminatory, multilateral and internationally and effectively verifiable treaty banning the production of fissile material for nuclear weapons or other nuclear explosive devices

The General Assembly,

Recalling its resolutions 48/75 L of 16 December 1993, 53/77 I of 4 December 1998, 55/33 Y of 20 November 2000, 56/24 J of 29 November 2001, 57/80 of 22 November 2002 and 58/57 of 8 December 2003,

Convinced that a non-discriminatory, multilateral and internationally and effectively verifiable treaty banning the production of fissile material for nuclear weapons or other nuclear explosive devices would be a significant contribution to nuclear disarmament and nuclear non-proliferation,

Recalling the 1998 report of the Conference on Disarmament, in which, inter alia, the Conference recorded that, in proceeding to take a decision on this matter, that decision was without prejudice to any further decisions on the establishment of further subsidiary bodies under agenda item 1 and that intensive consultations would be pursued to seek the views of the members of the Conference on Disarmament on appropriate methods and approaches for dealing with agenda item 1, taking into consideration all proposals and views in that respect,[1]

1. *Recalls* the decision of the Conference on Disarmament[1] to establish, under item 1 of its agenda entitled "Cessation of the nuclear arms race and nuclear disarmament", an ad hoc committee which shall negotiate, on the basis of the report of the Special Coordinator[2] and the mandate contained therein, a non-discriminatory, multilateral and internationally and effectively verifiable treaty banning the production of fissile material for nuclear weapons or other nuclear explosive devices;

2. *Urges* the Conference on Disarmament to agree on a programme of work that includes the immediate commencement of negotiations on such a treaty.

66th plenary meeting
3 December 2004

---

[1] See *Official Records of the General Assembly, Fifty-third Session, Supplement No. 27* (A/53/27), para. 10.
[2] CD/1299.

F. A/RES/58/57

The Conference on Disarmament decision (CD/1547) of 11 August 1998 to establish, under item 1 of its agenda entitled "Cessation of the nuclear arms race and nuclear disarmament", an ad hoc committee to negotiate, on the basis of the report of the Special Coordinator (CD/1299) and the mandate contained therein, a non-discriminatory, multilateral and internationally and effectively verifiable treaty banning the production of fissile material for nuclear weapons or other nuclear explosive devices

The General Assembly,

Recalling its resolutions 48/75 L of 16 December 1993, 53/77 I of 4 December 1998, 55/33 Y of 20 November 2000, 56/24 J of 29 November 2001 and 57/80 of 22 November 2002,

Convinced that a non-discriminatory, multilateral and internationally and effectively verifiable treaty banning the production of fissile material for nuclear weapons or other nuclear explosive devices would be a significant contribution to nuclear disarmament and nuclear non-proliferation,

Recalling the 1998 report of the Conference on Disarmament, in which, inter alia, the Conference recorded that, in proceeding to take a decision on this matter, that decision was without prejudice to any further decisions on the establishment of further subsidiary bodies under agenda item 1 and that intensive consultations would be pursued to seek the views of the members of the Conference on Disarmament on appropriate methods and approaches for dealing with agenda item 1, taking into consideration all proposals and views in that respect,[1]

1. *Recalls* the decision of the Conference on Disarmament[1] to establish, under item 1 of its agenda entitled "Cessation of the nuclear arms race and nuclear disarmament", an ad hoc committee which shall negotiate, on the basis of the report of the Special Coordinator[2] and the mandate contained therein, a non-discriminatory, multilateral and internationally and effectively verifiable treaty banning the production of fissile material for nuclear weapons or other nuclear explosive devices;

2. *Urges* the Conference on Disarmament to agree on a programme of work that includes the immediate commencement of negotiations on such a treaty.

71st plenary meeting
8 December 2003

[1] See *Official Records of the General Assembly, Fifty third Session, Supplement No. 27* (A/53/27), para. 10.

[2] CD/1299.

G. A/RES/57/80

The Conference on Disarmament decision (CD/1547) of 11 August 1998 to establish, under item 1 of its agenda entitled "Cessation of the nuclear arms race and nuclear disarmament", an ad hoc committee to negotiate, on the basis of the report of the Special Coordinator (CD/1299) and the mandate contained therein, a non-discriminatory, multilateral and internationally and effectively verifiable treaty banning the production of fissile material for nuclear weapons or other nuclear explosive devices

The General Assembly,

Recalling its resolutions 48/75 L of 16 December 1993, 53/77 I of 4 December 1998, 55/33 Y of 20 November 2000 and 56/24 J of 29 November 2001,

Convinced that a non-discriminatory, multilateral and internationally and effectively verifiable treaty banning the production of fissile material for nuclear weapons or other nuclear explosive devices would be a significant contribution to nuclear disarmament and nuclear non-proliferation,

Recalling the 1998 report of the Conference on Disarmament, in which, inter alia, the Conference records that, in proceeding to take a decision on this matter, that decision is without prejudice to any further decisions on the establishment of further subsidiary bodies under agenda item 1 and that intensive consultations will be pursued to seek the views of the members of the Conference on Disarmament on appropriate methods and approaches for dealing with agenda item 1, taking into consideration all proposals and views in that respect,[1]

1. *Recalls* the decision of the Conference on Disarmament[1] to establish, under item 1 of its agenda entitled "Cessation of the nuclear arms race and nuclear disarmament", an ad hoc committee which shall negotiate, on the basis of the report of the Special Coordinator[2] and the mandate contained therein, a non-discriminatory, multilateral and internationally and effectively verifiable treaty banning the production of fissile material for nuclear weapons or other nuclear explosive devices;

2. *Urges* the Conference on Disarmament to agree on a programme of work that includes the immediate commencement of negotiations on such a treaty.

57th plenary meeting
22 November 2002

[1] See *Official Records of the General Assembly, Fifty-third Session, Supplement No. 27* (A/53/27), para. 10.
[2] CD/1299.

H. A/RES/56/24 J The Conference on Disarmament decision (CD/1547) of 11 August 1998 to establish, under item 1 of its agenda entitled "Cessation of the nuclear arms race and nuclear disarmament", an ad hoc committee to negotiate, on the basis of the report of the Special Coordinator (CD/1299) and the mandate contained therein, a non-discriminatory, multilateral and internationally and effectively verifiable treaty banning the production of fissile material for nuclear weapons or other nuclear explosive devices

The General Assembly,

Recalling its resolutions 48/75 L of 16 December 1993, 53/77 I of 4 December 1998 and 55/33 Y of 20 November 2000,

Convinced that a non-discriminatory, multilateral and internationally and effectively verifiable treaty banning the production of fissile material for nuclear weapons or other nuclear explosive devices would be a significant contribution to nuclear disarmament and nuclear non-proliferation,

Recalling the 1998 report of the Conference on Disarmament, in which, inter alia, the Conference records that, in proceeding to take a decision on this matter, that decision is without prejudice to any further decisions on the establishment of further subsidiary bodies under agenda item 1 and that intensive consultations will be pursued to seek the views of the members of the Conference on Disarmament on appropriate methods and approaches for dealing with agenda item 1, taking into consideration all proposals and views in that respect,[1]

1. *Recalls* the decision of the Conference on Disarmament[1] to establish, under item 1 of its agenda entitled "Cessation of the nuclear arms race and nuclear disarmament", an ad hoc committee which shall negotiate, on the basis of the report of the Special Coordinator[2] and the mandate contained therein, a non-discriminatory, multilateral and internationally and effectively verifiable treaty banning the production of fissile material for nuclear weapons or other nuclear explosive devices;

2. *Urges* the Conference on Disarmament to agree on a programme of work that includes the immediate commencement of negotiations on such a treaty.

68th plenary meeting
29 November 2001

---

[1] See *Official Records of the General Assembly, Fifty-third Session, Supplement No. 27* (A/53/27), para. 10.
[2] CD/1299.

I. A/RES/55/33 Y
The Conference on Disarmament decision (CD/1547) of 11 August 1998 to establish, under item 1 of its agenda entitled "Cessation of the nuclear arms race and nuclear disarmament", an ad hoc committee to negotiate, on the basis of the report of the Special Coordinator (CD/1299) and the mandate contained therein, a non-discriminatory, multilateral and internationally and effectively verifiable treaty banning the production of fissile material for nuclear weapons or other nuclear explosive devices

The General Assembly,

Recalling its resolutions 48/75 L of 16 December 1993 and 53/77 I of 4 December 1998,

Convinced that a non-discriminatory, multilateral and internationally and effectively verifiable treaty banning the production of fissile material for nuclear weapons or other nuclear explosive devices would be a significant contribution to nuclear disarmament and nuclear non-proliferation,

Recalling the 1998 report of the Conference on Disarmament, in which, inter alia, the Conference records that, in proceeding to take a decision on this matter, that decision is without prejudice to any further decisions on the establishment of further subsidiary bodies under agenda item 1 and that intensive consultations will be pursued to seek the views of the members of the Conference on Disarmament on appropriate methods and approaches for dealing with agenda item 1, taking into consideration all proposals and views in that respect,[1]

1. *Recalls* the decision by the Conference on Disarmament[1] to establish, under item 1 of its agenda entitled "Cessation of the nuclear arms race and nuclear disarmament", an ad hoc committee which shall negotiate, on the basis of the report of the Special Coordinator[2] and the mandate contained therein, a non-discriminatory, multilateral and internationally and effectively verifiable treaty banning the production of fissile material for nuclear weapons or other nuclear explosive devices;

2. *Urges* the Conference on Disarmament to agree on a programme of work that includes the immediate commencement of negotiations on such a treaty.

69th plenary meeting
20 November 2000

[1] See *Official Records of the General Assembly, Fifty-third Session, Supplement No. 27* (A/53/27), para. 10.
[2] CD/1299.

J. A/RES/53/77 I
The Conference on Disarmament decision to establish, under item 1 of its agenda entitled "Cessation of the nuclear arms race and nuclear disarmament", an ad hoc committee to negotiate, on the basis of the report of the Special Coordinator (CD/1299) and the mandate contained therein, a non-discriminatory, multilateral and internationally and effectively verifiable treaty banning the production of fissile material for nuclear weapons or other nuclear explosive devices

The General Assembly,

Recalling its resolution 48/75 L of 16 December 1993,

Convinced that a non-discriminatory, multilateral and internationally and effectively verifiable treaty banning the production of fissile material for nuclear weapons or other nuclear explosive devices would be a significant contribution to nuclear non-proliferation in all its aspects,

Noting the report of the Conference on Disarmament, in which, inter alia, the Conference records that in proceeding to take a decision on this matter, that decision is without prejudice to any further decisions on the establishment of further subsidiary bodies under agenda item 1 and that intensive consultations will be pursued to seek the views of the members of the Conference on Disarmament on appropriate methods and approaches for dealing with agenda item 1, taking into consideration all proposals and views in that respect,[1]

1. *Welcomes* the decision by the Conference on Disarmament[1] to establish, under item 1 of its agenda entitled "Cessation of the nuclear arms race and nuclear disarmament", an ad hoc committee which shall negotiate, on the basis of the report of the Special Coordinator[2] and the mandate contained therein, a non-discriminatory, multilateral and internationally and effectively verifiable treaty banning the production of fissile material for nuclear weapons or other nuclear explosive devices;

2. *Notes with satisfaction* that this ad hoc committee has already engaged in the first step in the substantive negotiations;

3. *Encourages* the Conference on Disarmament to re-establish its ad hoc committee at the beginning of its 1999 session.

79th plenary meeting
4 December 1998

---

[1] See *Official Records of the General Assembly, Fifty-third Session, Supplement No. 27* (A/53/27), para. 10.
[2] CD/1299.

K. A/RES/48/75 L
Prohibition of the production of fissile material for nuclear weapons or other nuclear explosive devices

The General Assembly,

Welcoming the significant progress in reducing nuclear-weapon arsenals as evidenced by the substantive bilateral agreements between the Russian Federation and the United States of America and their respective unilateral undertakings regarding the disposition of fissile material,

Welcoming also the initiative of the United States of America concerning a multilateral, internationally and effectively verifiable treaty on the prohibition of the production of fissile material for nuclear weapons or other nuclear explosive devices,

Welcoming further the decision taken by the Conference on Disarmament on 10 August 1993 to give its Ad Hoc Committee on a Nuclear Test Ban a mandate to negotiate a universal and internationally and effectively verifiable comprehensive nuclear test-ban treaty,[1] and fully endorsing the contents of that decision,

Convinced that a non-discriminatory, multilateral and internationally and effectively verifiable treaty banning the production of fissile material for nuclear weapons or other nuclear explosive devices would be a significant contribution to nuclear non-proliferation in all its aspects,

1. *Recommends* the negotiation in the most appropriate international forum of a non-discriminatory, multilateral and internationally and effectively verifiable treaty banning the production of fissile material for nuclear weapons or other nuclear explosive devices;

2. *Requests* the International Atomic Energy Agency to provide assistance for examination of verification arrangements for such a treaty as required;

3. *Calls upon* all States to demonstrate their commitment to the objectives of a non-discriminatory, multilateral and internationally and effectively verifiable treaty banning the production of fissile material for nuclear weapons or other nuclear explosive devices;

4. *Decides* to include in the provisional agenda of its forty-ninth session an item entitled "Prohibition of the production of fissile material for nuclear weapons or other nuclear explosive devices".

81st plenary meeting,
16 December 1993

---

[1] See *Official Records of the General Assembly, Forty-eighth Session, Supplement No. 27* (A/48/27), para. 31 (para. 2 of the quoted text).

II. Report of Ambassador Shannon

CD/1299
Report of Ambassador Gerald E. Shannon of Canada on consultations on the most appropriate arrangement to negotiate a treaty banning the production of fissile material for nuclear weapons or other nuclear explosive devices

At the beginning of last year's session, I was tasked with seeking the views of members of the most appropriate arrangement to negotiate a non-discriminatory, multilateral and internationally and effectively verifiable treaty banning the production of fissile material for nuclear weapons or other nuclear explosive devices.

As you know, I held numerous consultations, both bilaterally and with groups and reported formally to this plenary on five occasions in 1994. Mid-way through the last session, consensus was reached that the CD was the appropriate forum to negotiate a treaty on this issue. At the end of the session in September, while there was no agreement on a mandate for an ad hoc committee, there was agreement, in principle, that an ad hoc committee be established on this issue as soon as a mandate had been agreed. At that time, the CD asked me to continue consultations on an appropriate mandate for an ad hoc committee in order to enable the convening of this Ad Hoc Committee as soon as possible.

At the beginning of this year's session, the Conference decided to continue consultations on a mandate.

I have since held numerous consultations, and am pleased to report that delegations have agreed that the mandate for such a committee should be based on resolution 48/75L of the United Nations General Assembly, and reads as follows:

1. *The Conference on Disarmament decides to establish an ad hoc committee on a "ban on the production of fissile material for nuclear weapons or other nuclear explosive devices".*

2. *The Conference directs the Ad Hoc Committee to negotiate a non-discriminatory, multilateral and internationally and effectively verifiable treaty banning the production of fissile material for nuclear weapons or other nuclear explosive devices.*

3. *The Ad Hoc Committee will report to the Conference on Disarmament on the progress of its work before the conclusion of the 1995 session.*

During the course of my consultations, many delegations expressed concerns about a variety of issues relating to fissile material, including the appropriate scope of the Convention. Some delegations expressed the view that this mandate would permit consideration in the Committee only of the future production of fissile material. Other delegations were of the view that the mandate would permit consideration not only of future but also of past production. Still others were of the view that consideration should not only relate to production of fissile material (past or future) but also to other issues, such as the management of such material.

It has been agreed by delegations that the mandate for the establishment of the Ad Hoc Committee does not preclude any delegation from raising for consideration in the Ad Hoc Committee any of the above noted issues.

Delegations with strong views were able to join consensus so we could all move forward on this issue. This means that an ad hoc committee on cut off can be established and negotiations can begin on this important topic. This has for some time been the common objective of all delegations of this Conference.

I have appreciated the productive contribution and support of all delegations in arriving at this result.

III. Views of Member States

A/68/154 and A/68/154/Add.1
Treaty banning the production of fissile material for nuclear weapons or other nuclear explosive devices

Report of the Secretary-General

I. Introduction

1. In its resolution 48/75 L of 16 December 1993, the General Assembly recommended the negotiation in the most appropriate international forum of a non-discriminatory, multilateral and internationally and effectively verifiable treaty banning the production of fissile material for nuclear weapons or other nuclear explosive devices, and called upon all States to demonstrate their commitment to the objectives of such a treaty.

2. In accordance with the resolution, the Conference on Disarmament initiated consultations among its members on what would be the most appropriate forum to negotiate such a treaty and what should be the mandate for negotiations. On 23 March 1995, the Conference endorsed the report of those consultations, calling for the establishment of an ad hoc committee in the Conference on Disarmament with the mandate to negotiate a treaty banning the production of fissile material for nuclear weapons or other nuclear explosive devices. In 1998, the committee met for negotiations for the three final weeks of the annual session of the Conference, but was not re-established the following year. Over the following 15 years, the Conference made several unsuccessful attempts to agree on and implement a substantive programme of work that would include negotiation of a fissile material cut-off treaty.

3. By paragraph 2 of resolution 67/53, the General Assembly requested the Secretary-General to seek the views of Member States on a treaty banning the production of fissile material for nuclear weapons or other nuclear explosive devices, including aspects thereof, and to submit a report on the subject to it at its sixty-eighth session. The present report is submitted pursuant to that request.

4. On 31 January 2013, the Office for Disarmament Affairs sent a note verbale to all Member States seeking their views. As at the time of writing of the present report, 37 Member States and the European Union had submitted their views, executive summaries of which are reproduced in section II. The reply of the European Union is reproduced in section III, in accordance with the modalities set out in resolution 65/276. Additional submissions will be issued as an addendum to the present report. The full texts of national submissions are available from http://www.unog.ch/80256EE600585943/(httpPages)/B8A3B48A3FB7185EC1257B28004 5DBE3?OpenDocument.

II. Replies received from Governments

Algeria

[Original: French]
[15 May 2013]

The treaty banning the production of fissile material for nuclear weapons or other nuclear explosive devices should be based on the parameters defined in resolution 48/75 L and the elements identified in the Shannon report (CD/1299), namely, a non-discriminatory, multilateral and internationally and effectively verifiable treaty.

This instrument should ban the future production of fissile material for nuclear weapons or other nuclear explosive devices and serve as a framework to reduce and eliminate stockpiles of such material.

The treaty should not be interpreted as infringing the inalienable right of States parties to pursue research on and the production and peaceful use of nuclear energy without discrimination.

The treaty should ban the future production of fissile material used in the manufacture of nuclear weapons or other nuclear explosive devices (highly enriched uranium and plutonium); the diversion of fissile material to purposes that are prohibited or contrary to the treaty objectives; the reactivation of decommissioned, dismantled or converted nuclear facilities for activities banned by the treaty; and the transfer of fissile material for the manufacture of nuclear weapons to other States or entities. It should also require States parties to decommission, dismantle and convert facilities that produce fissile material covered by the treaty and to reduce and eliminate their stockpiles of fissile material for the manufacture of nuclear weapons before the treaty enters into force.

The definitions of nuclear material contained in article XX of the Statute of the International Atomic Energy Agency (IAEA) could provide the basis for the treaty, with adjustments to be made as necessary.

The treaty should be underpinned by a rigorous and effective verification regime able to provide the necessary assurances of compliance with its provisions. To that end, the scope of the verification regime should be broad enough to cover the entire nuclear fuel cycle. It could be based on the comprehensive IAEA safeguards (INFCIRC/153 (Corrected)) that apply to non-nuclear-weapon States, and all military and civil nuclear activities during processing, use and storage should be placed under this regime.

By virtue of its mandate and experience, IAEA would be the appropriate agency to verify compliance with the treaty.

An effective safeguards regime requires that a sufficient amount of information and detailed inventories on the production, volume of fissile material stock for civil and military purposes, and data on civil and military nuclear facilities be made available.

The treaty should include measures to be invoked in the event of violations that entail non-compliance with fundamental treaty obligations, and measures to correct those situations and to resolve disputes that may arise among States parties concerning the application or interpretation of its provisions.

The duration of the treaty should be indefinite and should not be subject to reservations. The Secretary-General of the United Nations should be the depositary of the treaty.

Australia

[Original: English]
[15 May 2013]

An effectively verifiable treaty banning the production of fissile material for nuclear weapons or other nuclear explosive devices has the potential to deliver substantial benefits for the security of all States, furthering the twin goals of nuclear disarmament and nuclear non-proliferation.

Australia sees the Group of Governmental Experts established by General Assembly resolution 67/53 as a valuable opportunity. Technical work to develop practical elements for a treaty can be advanced without touching on broader political issues. This should be the case, for example, for many aspects of verification of a treaty.

Australia considers that the key undertaking of States parties to such a treaty should be not to produce fissile material for nuclear weapons or other nuclear explosive devices. Parties would also dismantle or convert to other uses facilities formerly used to produce fissile material for nuclear weapons. Australia would support the inclusion in the treaty of provisions on stocks of fissile material, whether produced for civil or military purposes. Such provisions should address the status of pre-existing stocks, as well as mechanisms under which States could choose to submit excess military stocks to irreversible peaceful use and verification commitments.

Fissile materials whose production would be controlled by the treaty should be those relevant to the manufacture of nuclear weapons or other nuclear explosive devices. A good basis for discussion of what constitutes such material is the definition of "direct-use materials" used by the International Atomic Energy Agency (IAEA).

Australia considers that cost-effective verification of the treaty could best be achieved by focusing on facilities for enrichment of uranium and for separation of plutonium, including those formerly used for nuclear weapons purposes, and on facilities processing or using fissile material subject to the treaty. We recognize nevertheless that it may also be necessary to gain assurance of compliance through declaration and inspection of some other nuclear material and activities. Verification against undeclared production of relevant fissile material would need to be an essential element of the treaty. Australia notes that most of the verification requirements of a treaty could already be met for many States parties by the application of comprehensive IAEA safeguards together with an additional protocol that meets the requirements of INFCIRC/540 (corrected).

Austria

[Original: English]
[14 May 2013]

Austria is in favour of the speedy conclusion of a non-discriminatory, multilateral and internationally and effectively verifiable treaty banning the production of fissile material for nuclear weapons or other nuclear explosive devices to accelerate progress towards a nuclear-weapon-free world.

Brazil

[Original: English]
[22 May 2013]

A fissile material treaty should contribute effectively to the achievement of the goal of a world without nuclear weapons. Given the amount of highly enriched uranium and plutonium already accumulated by States possessing nuclear weapons, a treaty would bring added value to nuclear disarmament only if it contained specific commitments related to past production.

The treaty should address both future and past production. It is clear that future production of fissile material for nuclear weapons and other explosive devices should be prohibited outright. With regard to past production, it will be one of the tasks of the Group of Governmental Experts to explore possible options, including a phased process of destruction of all pre-existing weapons-grade fissile material.

In 2010, Brazil proposed to the Conference on Disarmament, in a conceptual working paper (CD/1888), a possible general structure for a treaty on fissile material for nuclear weapons or other nuclear explosive devices, comprising a framework or umbrella treaty and two protocols, with their respective verification mechanisms.

With comprehensive safeguards agreements, non-nuclear-weapon States already have in place commitments tantamount to such a treaty. The purpose of the agreements entered into by these States pursuant to article III of the Treaty on the Non-Proliferation of Nuclear Weapons is precisely to ensure that nuclear material will not be used for nuclear weapons or other nuclear explosive devices.

With regard to nuclear-weapon States, the treaty would have to provide for the application of IAEA safeguards in all facilities containing fissile material, with the exception of the material already present in weapons systems themselves.

Canada

[Original: English]
[15 May 2013]

Canada accords the utmost importance to beginning negotiations on a treaty banning the production of fissile material for use in nuclear weapons. Such a treaty is not an end unto itself but a concrete and practical step towards nuclear non-proliferation and disarmament.

Any definition of fissile material used in a future treaty must be broad enough to ensure that all fissile material relevant to nuclear weapons or other nuclear explosive devices, now or in the future, is captured. Wherever possible, Canada is in favour of using relevant IAEA terminology and definitions.

At a minimum, the scope of a future treaty should ban the production of fissile material for nuclear weapons and other nuclear explosive devices. Verification of additional fissile material, both civilian and military, should also be considered, given the risk of diversion to weapons use, including through the use of a mechanism similar to existing comprehensive safeguards agreements.

Existing stockpiles of fissile material remain a difficult and contentious issue which will likely only be resolved through direct negotiation. Canada remains committed to the Shannon mandate (CD/1299), and does not preclude additional measures to address existing stockpiles, in or alongside the framework of a treaty, such as increasing transparency; declaring excess fissile material and placing it under verification; and disposing of excess fissile material.

It is possible to develop technically, financially, legally and politically effective verification measures by using, extending or adapting elements of the IAEA comprehensive safeguards regime, including additional measures for existing stockpiles and declared excess fissile material. We envision a key role for IAEA, including as the potential verifying agency. This could be cost-effective, although potential broader implications for the Agency, including financial, would have to be carefully considered. Other measures in the treaty could include bilateral and multilateral verification measures, and national technical means.

Consideration could also be given to a mechanism by which to review compliance and treaty implementation, such as a review process; an IAEA secretariat; or a separate secretariat to address compliance and implementation.

China

[Original: Chinese, English]
[14 May 2013]

The Government of China is of the view that conclusion of a treaty banning the production of fissile material for nuclear weapons or other nuclear explosive devices through negotiation is an important issue in the international arms control process. The complete ban on the production of fissile material for nuclear weapons or other nuclear explosive devices through conclusion of the treaty, will be conducive to promoting nuclear disarmament and non-proliferation, and is an important step leading to the complete prohibition and thorough destruction of nuclear weapons.

China has always stood for concluding a non-discriminatory, multilateral and internationally effectively verifiable treaty as early as possible in the Conference on Disarmament on the basis of document CD/1299 and the mandate contained therein. As the sole multilateral disarmament negotiating forum, the Conference is the only appropriate venue for the negotiation of such a treaty. China supports the Conference towards agreement on a comprehensive and balanced programme of work so as to carry out substantive works, including negotiation of the treaty.

The Government of China believes that the Group of Governmental Experts on the treaty should include all States with the ability to produce fissile material for nuclear weapons or other nuclear explosive devices, and it should take full consideration of equitable geographical representation. The Group should operate within the mandate of General Assembly resolution 67/53. It should discuss relevant issues and make recommendations according to the principle of consensus. Should the Conference agree upon and implement a programme of work, the Group should conclude, and its work should be transmitted to the Conference.

The Government of China is of the view that the treaty should mainly contain such aspects as definitions, obligations, organization, verification and entry into force. The treaty should cover only the future production of fissile material for nuclear weapons or other nuclear explosive devices.

The Government of China has been actively supporting the commencement of the negotiation of the treaty. China has joined the consensus or voted in favour of resolution 48/75 in 1993 and all General Assembly resolutions related to the treaty since then. China took part in the adoption of the Final Document of the 2010 Review Conference of the Parties to the Treaty on the Non-Proliferation of Nuclear Weapons by consensus and has been actively promoting the implementation of the action plan related to the treaty banning the production of fissile material for nuclear weapons or other nuclear explosive devices proposed by the Final Document. China supports the decision of the Conference on the mandate, working mechanism and relevant issues of the treaty negotiation. It has played a serious part in all treaty-related activities within the framework of the Conference. China will continue its efforts to promote an early commencement of negotiation of the treaty in the Conference.

Congo

[Original: French]
[15 May 2013]

The concept of the treaty banning the production of fissile material for nuclear weapons or other nuclear explosive devices is an opportunity to revitalize the Conference on Disarmament and to take a leap forward in the disarmament process.

Nevertheless, it is worth considering the view of the United States of America that, with respect to both the Treaty on the Non-Proliferation of Nuclear Weapons and this treaty, the deadlock is political in nature. Should the Member States reach a global consensus at the political level, a solution would be found to the other legal and technical issues.

A process should therefore be envisaged that will take into account the different positions of Member and non-Member States, nuclear-weapon and non-nuclear-weapon States, and the group of countries that includes India, Israel, Pakistan, Egypt, the Islamic Republic of Iran and the Democratic People's Republic of Korea on specific issues that are often the subject of misgivings, such as transparency on the quantity and quality of fissile material stocks, even for civil purposes, and the role IAEA should play in that regard.

Consequently, the Congo supports the proposal, put forward by Brazil, to establish a group of governmental experts in order to expedite the process on the basis of the views of Member States which would propose an agenda and recommendations to the Secretary-General.

With regard to the treaty, its scope should include not only a ban on the production of fissile material for military purposes, but also a limit on the levels of fissile material stocks for civil purposes. It should also cover production and enrichment technology as well as the relevant equipment and facilities.

Moreover, it is clear that components of "fissile material" and everything related to it must be defined. In addition, it would be impossible to draft a treaty without determining its scope, the timetable for entry into force and verification mechanisms.

As with the Treaty on the Non-Proliferation of Nuclear Weapons, the scope of the treaty would be unlimited, since the goal is a world free of nuclear weapons. Production for military purposes must therefore be brought to a halt and new production and the development of production techniques must not be allowed. In that regard, IAEA and the Comprehensive Nuclear-Test-Ban Treaty Organization are tools that could make a useful contribution to the verification and monitoring mechanism. For this treaty to be effective its scope must be universal.

Cuba

[Original: Spanish]
[20 May 2013]

Cuba favours the commencement of multilateral negotiations on a non-discriminatory, multilateral and internationally and effectively verifiable treaty, which will ban the production of fissile material for nuclear weapons or other nuclear explosive devices and will also address the question of existing stocks. A fissile material treaty should not focus solely on non-proliferation, but should be an instrument for nuclear disarmament. That is why we believe that the treaty is a step towards the prime objective of nuclear disarmament. The so-called Shannon mandate is still fit for purpose and sufficiently broad in scope to cover the interests of all delegations. Once the relevant subsidiary body is established within the Conference on Disarmament to start negotiations, all Member States must be free to take up and address all issues they consider to be relevant to the future treaty.

The main purpose of the treaty must be a ban on the production of fissile material for nuclear weapons and it should also cover fissile material that existed before the treaty's entry into force.

The provisions of a future fissile material treaty must not interfere with the right to use such material for peaceful purposes. The future treaty must also promote international cooperation on the peaceful use of nuclear energy.

The IAEA criteria should be used to define the concept of fissile material. To ensure that the treaty will be an effective instrument, it should be governed by the fundamental principles of transparency, verification and irreversibility.

The future treaty must have an effective verification mechanism that will also ensure irreversibility. The IAEA safeguards system could be used as a tool for the implementation of the verification mechanism.

The future treaty should be negotiated within the Conference on Disarmament, the only multilateral forum for the negotiation of disarmament treaties. We therefore reiterate the need for the Conference on Disarmament to adopt as soon as possible a balanced and comprehensive programme of work that takes into account the real disarmament priorities.

For Cuba, nuclear disarmament is and must remain the highest priority in the field of disarmament and should be given the highest priority on the Conference's programme of work.

Ecuador*

[Original: Spanish]
[13 September 2013]

The Constitution of Ecuador declares peace and universal disarmament to be principles of international law, establishes the concept of living well and grants rights to nature. This domestic legislation, as well as the State obligations emanating from the Treaty on the Non-Proliferation of Nuclear Weapons and the Treaty for the Prohibition of Nuclear Weapons in Latin America and the Caribbean (Treaty of Tlatelolco), have made Ecuador a country with an eminently pacifist

* The reply from Ecuador was issued as an addendum to the main report with the symbol A/68/154/Add.1.

and humanitarian mission, which envisions a world in which weapons of mass destruction have no place.

Ecuador believes that the issues affecting international peace and security, in particular matters relating to nuclear disarmament and nuclear non-proliferation, must be discussed and addressed by all States, and not merely a few States.

In its resolution 67/53, the General Assembly provides that, in 2014 and 2015, the group of governmental experts will make recommendations on possible aspects that could contribute to but not negotiate a treaty banning the production of fissile material, taking into account the views and requirements of Member States. Ecuador considers that those recommendations should be based on document CD/1299 and the mandate contained therein (the Shannon mandate).

Ecuador believes that nuclear disarmament remains the outstanding task: more than 40 years since they entered into force, the provisions of article VI of the Treaty on the Non-Proliferation of Nuclear Weapons have yet to be implemented. Therefore, for Ecuador, as well as for the Movement of Non-Aligned Countries, progress in the area of nuclear disarmament is the priority, as affirmed in article VI of the Treaty and in the agreements resulting from the 2000 and 2010 review conferences of the parties to the Treaty. In that regard, Ecuador has respected and will continue to respect its commitments in the area of nuclear non-proliferation.

For the above reasons, Ecuador believes it is critical that a treaty banning the production of fissile material should approach the issue from the perspectives of nuclear disarmament and nuclear non-proliferation. This implies that negotiations on the treaty should address both existing stocks and future production. A treaty whose scope only covered future production and excluded stocks would be incomplete.

Ecuador would be willing to support a "cut-off treaty", as it fully subscribes to the principles that underpin nuclear non-proliferation. However, it should be noted that a treaty of this kind would not cover existing stocks of fissile material. In view of the foregoing, Ecuador considers that, from the beginning, this treaty should not be called a "cut-off treaty", as this title would immediately limit the scope and content of the treaty.

As stipulated in resolution 67/53, the group of governmental experts will not negotiate, but instead will formulate recommendations. It follows that the Conference on Disarmament, as the only negotiating forum on nuclear disarmament, is still the only forum for negotiation of a treaty banning the production of fissile material. Such negotiation should therefore be conducted at the Conference. However, the existence and continuation of the Conference cannot and should not depend on negotiations on this treaty.

As is clear from document CD/1864, the question of a treaty banning the production of fissile material is just one of various pending matters for the Conference and the nuclear disarmament agenda. The urgent need to initiate negotiations on negative security assurances, a nuclear weapons convention and the peaceful use of outer space must also be considered. For States that do not have nuclear weapons, those issues are as important or more important than a treaty banning the production of fissile material.

The inalienable right of States to use nuclear energy for peaceful purposes, which constitutes the third pillar of the Treaty on the Non-Proliferation of Nuclear Weapons, must be fully considered and respected during any discussion and negotiation process on a treaty banning the production of fissile material.

Finland

[Original: English]
[13 May 2013]

The amount of fissile material for weapons purposes in the world should be as small as possible, and the existing fissile material stockpiles should be subject to as strict control as practicable.

A treaty banning the production of fissile material for nuclear weapons or other nuclear explosive devices must not in any way form a barrier to peaceful uses of nuclear energy.

Finland supports the start of treaty negotiations as soon as possible. Finland is in favour of a multilateral, non-discriminatory and verifiable treaty. The treaty would be an important next step in the pursuit of disarmament and the goals of the Treaty on the Non-Proliferation of Nuclear Weapons.

The Group of Governmental Experts should discuss the most controversial issues. One of the Group's main tasks would be to consider the verification mechanism of such a treaty.

France

[Original: French, English]
[14 May 2013]

The launch of negotiations at the Conference on Disarmament on a treaty banning the production of fissile material for nuclear weapons or other nuclear explosive devices is a priority for France. It is the next logical step towards creating the conditions for a world free of nuclear weapons, in accordance with the goals of the Treaty on the Non-Proliferation of Nuclear Weapons within the framework of a realistic approach based on gradual concrete gestures. France considers resolution 67/53 to be a useful contribution to discussions with a view to preparing the future negotiation of a legally binding international instrument.

The goal of the treaty is to quantitatively limit arsenals through the shutdown of production of fissile material for nuclear weapons. Like the Comprehensive Nuclear-Test-Ban Treaty, which it supplements, the fissile material treaty should be a universal treaty. As its aim is to contribute to nuclear disarmament, it appears essential that all countries which currently possess nuclear weapons should accede to it. Regarding non-nuclear-weapon States parties to the Non-Proliferation Treaty, and insofar as the fissile material treaty does not primarily aim to combat nuclear proliferation, it should involve no additional obligation for these countries. By the time it enters into force, all States concerned should declare, as France has already done, a moratorium on the production of fissile material for nuclear weapons.

The scope of the fissile material treaty is determined by the goal which is sought. It should cover only material and facilities which could truly allow circumvention of the treaty. As the treaty covers the shutdown of the production of fissile material for nuclear weapons, the corresponding preconstituted stocks are by definition excluded from its scope. Moreover the treaty should not ban the production of fissile material for civilian uses or for non-explosive military purposes.

The implementation of the treaty cannot be dissociated from the establishment of a credible verification regime. Verification must be carried out in compliance with two fundamental principles: preserving national security interests, and preventing the transfer of confidential information on nuclear weapons, in accordance with article I of the Non-Proliferation Treaty. Verification has three chief goals:

- Certifying the shutdown of dedicated means of production until dismantled or converted for civilian purposes
- Preventing, through the controls of IAEA in the relevant civilian facilities with regard to the goal of the treaty, the diversion of fissile material from civilian activities
- Handling suspected prohibited activities.

Regarding diplomatic provisions, France states its preference for the conclusion of a treaty of unlimited duration. It should also provide for a system allowing effective sanctioning of violations, and a withdrawal clause whose conditions must be regulated to avoid any abusive exercise of that right.

Germany

[Original: English]
[9 May 2013]

Germany attaches fundamental importance to the early commencement of negotiations on a treaty banning the production of fissile material, without prejudice to its actual scope, as the next logical and sensible step towards nuclear disarmament. Germany also considers such a treaty an important non-proliferation instrument. The treaty would offer a unique opportunity to establish a non-discriminatory treaty regime by creating essentially equal obligations for nuclear-weapon possessor States and non-nuclear-weapon States alike; cap the quantitative nuclear arms race and thereby instil new momentum into the process of nuclear disarmament, thus helping to maintain the integrity of the Treaty on the Non-Proliferation of Nuclear Weapons. Therefore, Germany believes that blocking the commencement of negotiations on the treaty banning the production of fissile material for nuclear weapons or other nuclear explosive devices, which condemns the world's sole multilateral disarmament negotiating forum, the Conference on Disarmament, to further stalemate, is not in the well-understood long-term interest of any State.

Hungary

[Original: English]
[9 May 2013]

In the view of Hungary, the elimination of nuclear weapons is not a single act, but rather a step-by-step process, in which the banning of the production of fissionable materials for weapons purposes is the long overdue "next logical step". The early commencement of negotiations on a treaty banning the production of fissile material for nuclear weapons or other nuclear explosive devices is an internationally recognized priority, which has been reaffirmed by important decisions and documents of different multilateral forums.

The Hungarian proposal includes suggestions for the following:

- A definition of fissile material
- A definition of what the production of fissile material entails

- A verification system entrusted to IAEA
- The inclusion of existing stockpiles in the scope of a fissile material treaty.

Definition. Hungary suggests that the definition of "fissile material" should contain neptunium-237, plutonium-239, plutonium mixtures, uranium-233, uranium enriched in the isotope 235, with the exception of plutonium mixtures with plutonium-238 concentrations equal to or more than 80 per cent; uranium enriched in the isotope 235 with concentrations less than 20 per cent; and fissile materials mixed with fission products (irradiated).

Production. Hungary suggests that "production" includes the following: (a) enrichment of uranium in uranium-235; (b) separation of plutonium and/or neptunium-237 from irradiated uranium; (c) separation of uranium-233 from irradiated thorium; and (d) conversion of fissile material into weapon-usable form.

Verification. Hungary is in favour of a mandate for the treaty verification being given to IAEA.

Scope/stockpiles. The elimination of existing stockpiles is envisaged to be part of the treaty on a phase-out basis. The issue of stockpiles should not hamper the verification of a fissile material cut-off right after the treaty's entry into force. Establishment of an accountancy and verification system of fissile material stockpiles may be subject to an additional protocol to the treaty at a later stage.

India

[Original: English]
[14 May 2013]

Without prejudice to the priority attached to nuclear disarmament, India supports international efforts aimed at early commencement of negotiations on a treaty banning the production of fissile material for nuclear weapons or other nuclear explosive devices in the Conference on Disarmament, in accordance with the mandate explicitly reflected in resolution 48/75 L, and later reconfirmed in the Shannon report (CD/1299), to negotiate a non-discriminatory, multilateral and internationally and effectively verifiable treaty banning the production of fissile material for nuclear weapons or other nuclear explosive devices. The agreed mandate set out in CD/1299 continues to be valid and relevant and should remain unchanged.

The Conference as the world's single multilateral disarmament negotiating body is the appropriate forum for negotiating such a treaty. The Conference as the agreed forum for treaty negotiations was one of the key elements of the consensus contained in CD/1299. The work on the treaty in the Conference and its subsidiary body should be conducted in accordance with the rules of procedure of the Conference and on the basis of strict adherence to the rule of consensus.

A fissile material treaty must be a treaty for banning the future production of fissile material for nuclear weapons or other nuclear explosive devices. If implemented in good faith through universal participation and adherence, it will make a significant contribution to nuclear non-proliferation in all its aspects. It would be a step towards nuclear disarmament but would not in itself be a disarmament measure.

The obligations and responsibilities arising from the treaty must apply in a non-discriminatory manner in particular, to all States parties directly affected by the treaty's obligations and responsibilities. The treaty would be global in character, thus excluding any regional specificity. It should include all States which are essential stakeholders for the treaty and thus critical for its universal adherence. The dynamic correlation between scope, definitions and verification will be an important factor in the treaty, also taking into account the costs of implementing the treaty. The mechanism for verifying the obligations enshrined in the treaty will be decided in the treaty negotiations and cannot be prejudged or agreed in advance. The treaty should not place an undue burden on military non-proscribed activities.

India would be willing to join only a non-discriminatory, multilaterally negotiated and internationally verifiable treaty, as and when it is concluded in the Conference, provided its national security interests are fully addressed. India is a nuclear-weapon State and a responsible member of the international community and will approach treaty negotiations as such. The establishment of the Group of Governmental Experts under General Assembly resolution 67/53 cannot and does not replace the Conference as the forum for the negotiation of the treaty. The work of the proposed Group amounts to neither prenegotiations nor negotiations on a treaty, which should take place in the Conference.

Indonesia

[Original: English]
[16 May 2013]

Global nuclear disarmament is the highest priority, and any substantive advancement on a fissile material treaty should be in keeping with the objective of achieving the complete elimination of all nuclear weapons.

While we agree that there should be negotiations for the treaty, we remain of the view that there should also be negotiations for a nuclear weapons convention, negative security assurances and the prevention of an arms race in outer space.

The fissile material treaty must ban existing stocks as well as future production of fissile materials for nuclear weapons or other nuclear explosive devices. Additionally, the scope of the treaty should extend to an effective and transparent international verification mechanism.

The treaty should be negotiated in the Conference on Disarmament, the sole multilateral negotiating forum for disarmament. Furthermore, it should be multilateral, non-discriminatory and verifiable.

International verification of such a treaty will make a vital contribution to global nuclear disarmament, and create confidence and trust for its possible universality. Many non-nuclear-weapon States have accepted the IAEA comprehensive safeguards on all of their nuclear programmes in order to verify their commitment to not divert nuclear materials for weapons. However, nuclear-weapon States are not required to have similar safeguards on their nuclear facilities. A verified fissile material treaty will help to redress the imbalance between the rights and obligations of nuclear-weapon States and non-nuclear-weapon States. The International Atomic Energy Agency can be given the task of conducting the verification mechanism for the treaty.

The treaty must clearly state that the States parties, in addition to eliminating their existing stocks, shall not produce, acquire from any source or transfer to any recipient or use fissile materials for nuclear weapons or other nuclear explosive devices.

The object and purpose of a fissile material treaty must reflect the treaty as a concrete tool in achieving the goal of a world without nuclear weapons.

Islamic Republic of Iran

[Original: English]
[30 May 2013]

In the view of the Islamic Republic of Iran, the existence of nuclear weapons and their vertical and horizontal proliferation is the most serious threat to international peace and security. Therefore, the total elimination of nuclear weapons is the only absolute guarantee against their threat and use, and a prerequisite for the establishment of a nuclear-weapon-free world. The first and best practical measure to achieve this noble goal is the early conclusion of a nuclear weapons convention to prohibit the production, development, possession, stockpiling, use or threat of use of nuclear weapons under any circumstances, and provide for their total elimination at the earliest date, in an irreversible and transparent manner and under strict international verification.

The Islamic Republic of Iran strongly believes that any instrument which is to ban the production, and provide for the total elimination, of fissile material for nuclear weapons or other nuclear explosive devices, should be comprehensive, non-discriminatory and under strict international verification. Accordingly, its scope must cover the past, present and future production of fissile material and provide for the declaration and total elimination of all stocks of such material at a fixed date, in an irreversible and transparent manner and under strict international verification. It also should not provide, in any way whatsoever, the grounds for the recognition of any new status for nuclear-weapon possessors. It should oblige all nuclear-weapon possessors and all nuclear-weapon States, without exception, to completely end the production of fissile material and to declare and destroy all their stockpiles of such material within a specified framework of time, in an irreversible, transparent and internationally verifiable manner. Additionally, such an instrument must be of a nuclear disarmament nature and not another non-proliferation instrument, and should not add to the burden of the non-nuclear-weapon States parties to the Treaty on the Non-Proliferation of Nuclear Weapons.

Ireland

[Original: English]
[17 May 2013]

To constitute an effective measure relating to nuclear disarmament, within the terms of article VI of the Treaty on the Non-Proliferation of Nuclear Weapons, Ireland suggests that the objectives of the treaty banning the production of fissile material for nuclear weapons or other nuclear explosive devices should be to prohibit any future production of fissile material for nuclear weapons or other nuclear explosive devices; to require the placement of all civil stocks of fissile material under verification safeguards, such that the material cannot be diverted to uses in nuclear weapons or other nuclear explosive devices; to require the placement of all non-civil stocks declared as excess under verification safeguards, such that the material cannot be diverted to uses in nuclear weapons or other nuclear explosive devices; and to require the destruction or conversion to exclusively peaceful uses of any facilities previously used for the production of fissile material for nuclear weapons or other nuclear explosive devices. These core objectives should form the basis of the treaty's general obligations.

In terms of defining the term "fissile material", the starting point should be the definition contained in article XX of the Statute of IAEA. The Agency should be given the task of advising the Group of Governmental Experts on related issues.

The Agency should also be invited to advise the Group on the feasibility of using existing safeguards structures and arrangements as a means to aid the implementation of the treaty, and to offer views as to whether and to what extent new structures or arrangements might require to be devised, having regard to the particular objectives of the treaty. To that end, useful guidance might also be sought from other treaty regimes, such as, for example, the Chemical Weapons Convention and the Comprehensive Nuclear-Test-Ban Treaty.

Italy

[Original: English]
[13 May 2013]

A substantive discussion on the treaty banning the production of fissile material for nuclear weapons or other nuclear explosive devices should address five issues, which are set out below.

Definitions. Two key principles should apply: feasibility and credibility. The definition of fissile materials should include those materials that with current or near-term predictable technology and equipment have a reasonable probability of being used in the manufacture of nuclear explosive devices, principally the relevant isotopes of uranium and plutonium. Furthermore, the definition of fissile materials should be such that it allows verification without undue technical complications or excessive expenditure. Any definition should be broad enough to make the treaty credible and effective, but not so extensive as to imply unacceptably complex and expensive verification procedures or unnecessary limits to the peaceful uses of nuclear energy.

Verification. Provisions on verification are essential to any disarmament and non-proliferation treaty. A treaty verification system should aim at both fissile materials and production facilities. As for fissile materials, the purpose should be to verify any discrepancy between actual production and the declared one and the non-diversion of existing fissile material, including that in civil use. With regard to production facilities, it should be to ascertain the absence of undeclared production and the irreversible conversion or dismantlement of production facilities formerly used for nuclear weapons purposes. The agency given the task of verification should be IAEA.

Nuclear fuel. Highly enriched fissile material is also used as fuel for naval vessels. Negotiation on a treaty will have to decide whether this fissile material should be covered by the provisions of the treaty, and, if so, how.

Production plants. The problem of production facilities of weapons-grade fissile material is to be addressed.

Stockpiles. Stockpiles are the real stumbling block. The views expressed so far in the Conference on Disarmament, unsurprisingly, have been radically different, but in this line of work nothing is ever black or white. Thinking it through should lead to possible compromise solutions.

Japan

[Original: English]
[13 May 2013]

Japan is of the view that a treaty banning the production of fissile material for nuclear weapons or other nuclear explosive devices:

(a) Should be negotiated in accordance with the mandate contained in the Shannon report (CD/1299). In view of the sustained consensual support for the mandate, its reopening would be counterproductive. Moreover, it provides a flexible basis for dealing with existing stocks in the course of negotiations (see paragraph 9 of the full report of Japan's views, available from http://www.unog.ch/80256EDD006B8954/(httpAssets)/778B94BEAE525FAFC1257B7C0041839D/$file/JAPAN.pdf);

(b) Should encompass various activities as core obligations rather than focus merely on a commitment not to undertake production of fissile material for nuclear weapons. The core obligations may include closing down/decommissioning former production facilities and refraining from reversion/diversion of fissile material from civil to military purposes (paragraphs 11-12);

(c) Should feature a broad general-purpose criterion to define prohibited materials and production activities ("definitions"), to the extent that this does not adversely impact peaceful uses of nuclear energy (paragraphs 14-15);

(d) Should provide for verification, not of only non-production of fissile material but of other obligations as well ("verification") (paragraphs 16-17);

(e) Can treat "definitions" and "verification" flexibly and separately. Technological limitations and considerations of cost-effectiveness should not restrict the scope of materials and activities subject to verification (paragraphs 18-19);

(f) Can address various aspects pertaining to existing stocks of fissile material. Regarding the matter of whether to include existing stocks within the scope of the treaty, it would be best to first examine concretely what is meant by "existing stocks" and "including within the scope" of the treaty. Only then will it be possible to find common ground on the issue (paragraphs 20-23).

Libya

[Original: Arabic]
[9 May 2013]

The States members of the League of Arab States reaffirm that the Conference on Disarmament is the sole multilateral negotiating body for disarmament operating under the auspices of the United Nations, in accordance with the mandate established by the first special session of the General Assembly devoted to disarmament, held in 1978. That session also affirmed that nuclear disarmament was an absolute priority.

The Group of Arab States emphasizes that the deadlock in the Conference on Disarmament is in no way due to any shortcoming in the Conference itself, but can rather be attributed to the lack among certain parties of the political will needed to make tangible progress towards disarmament and, consequently, the elimination of nuclear weapons.

The four items on the Conference agenda, namely, nuclear disarmament, a treaty banning the production of fissile material for nuclear weapons or other nuclear explosive devices,[1] prevention of an arms race in outer space and negative security assurances, are part of the comprehensive agenda on nuclear disarmament and no single item should not be given priority over the others.

In that connection, the Group of Arab States reiterates its appeal to the Conference on Disarmament to agree on a balanced and comprehensive programme of work that also includes the commencement of negotiations on a treaty banning the production of fissile material. The Group also reiterates its appeal for flexibility with a view to facilitating agreement on such a programme.

The Group of Arab States further believes that any negotiations on a treaty banning the production of fissile material must take place on the basis of the following:

The Conference on Disarmament must be the sole negotiating forum. All efforts should be aimed at facilitating negotiations under the auspices of the Conference, not establishing parallel mechanisms;

The treaty must be comprehensive, non-discriminatory and internationally verifiable;

In order to achieve the goal of disarmament, the treaty must prohibit the future production and stockpiling of such material.

Mexico

[Original: Spanish]
[17 May 2013]

Mexico, like other States parties to the Treaty on the Non-Proliferation of Nuclear Weapons, recognizes the need to conclude a treaty banning the production of fissile material for nuclear weapons or other nuclear explosive devices that includes existing fissile material, as one of the steps towards the complete elimination of nuclear weapons. The treaty negotiations should be part of a broad and comprehensive nuclear disarmament and non-proliferation process.

In that connection, the said instrument must contemplate the regulation of existing fissile material, a verification mechanism and confidence-building measures.

Mexico believes that a verification system must be defined to ensure that information is kept secure and confidential, under the IAEA verification system, without the need to create a new agency.

Bearing in mind that IAEA and several of its member States have laboratories and staff qualified to carry out the work of monitoring the source of fissile material, it would be unnecessary to create new infrastructure for that purpose.

[1] Without prejudice to the position of the Group of Arab States set out in this document, the name of the treaty is given as it appears in the resolution.

It is critical that fissile material storage facilities be placed under an accountability and control system with the broadest possible scope, since they are a proliferation risk. Mexico therefore suggests that States should declare all fissile material in their possession, in an effort to reduce reserves gradually, and that they should place that material under the IAEA comprehensive safeguards system.

It is also essential that the treaty contemplate banning the "direct use" of fissile material for the production of nuclear weapons and the transfer for nuclear weapon-related purposes of fissile material produced for civilian use. In addition, the future treaty should class neptunium and americium as fissile material, given their fissile nature and potential for use in a nuclear weapon.

In accordance with its obligations as a non-nuclear-weapon State and State party to the Treaty on the Non-Proliferation of Nuclear Weapons, Mexico supports promoting the use of low-enriched uranium for peaceful purposes and banning States not party to the treaty from acquiring or receiving fissile material for nuclear weapons or other nuclear devices and from helping third countries in any way to produce fissile material for explosives.

Netherlands

[Original: English]
[14 May 2013]

The Netherlands aims for an ambitious treaty banning the production of fissile material for nuclear weapons or other nuclear explosive devices. At the same time, we realize that reaching that aim requires mutual understanding of the political and technical challenges and a willingness to compromise, as there are many differing views on the fissile material treaty. We look forward to contributing to reaching a successful treaty in a creative, pragmatic and flexible way. We have outlined several ideas below.

Limiting the amount of weapons-grade nuclear material is consistent with our aim to enhance nuclear security and one of the priorities of the nuclear security summit which the Netherlands will host in 2014. In the absence of progress in the Conference on Disarmament and to ensure continued attention and improve the prospects for future negotiations, the Netherlands organized and funded several events and meetings on a fissile material treaty, aiming to identify possibilities and challenges for such negotiations.

In the view of the Netherlands, the basic elements of such a treaty could include a ban on producing fissile material for nuclear weapons or other nuclear explosive devices, a ban on acquiring fissile material for weapons and on transferring it to third countries, a ban on converting fissile material for civilian purposes for use in weapons and a requirement that States parties to the treaty disable, decommission and, where feasible, dismantle their fissile material production facilities for weapons, or reconfigure them.

The treaty should also contain provisions on an effective verification mechanism. It would be logical to give IAEA the task of verifying the obligations of States parties under such a treaty, as the Agency is the organization with the most capacity and international experience on nuclear safeguards. Verification should focus chiefly on existing and former military fissile material production facilities. Preferably, States possessing nuclear weapons should join the IAEA safeguards regime for their entire nuclear fuel cycle. Access to military facilities could be managed by provisions similar to the "managed access" clauses currently used to prevent sensitive commercial information from being disclosed during inspections of civilian nuclear fuel cycle processes. The Netherlands, hosting a significant peaceful nuclear industry, is happy to share its experiences with safeguarding and verifying nuclear facilities, including those relating to managed access.

With regard to the scope of the treaty—whether to include existing military stocks or not—the Netherlands would like to be ambitious. At the same time, as we are aware of the different points of view, we aim to work towards a compromise. One issue that has to be prominently addressed is transparency. Transparency should be both the result of a fissile material treaty and an important building block towards it. Further issues that need addressing are the costs of verification and the entry into force of the treaty.

We believe that the Shannon mandate is a good basis for negotiations, but flexibility from all sides is important concerning the exact wording of a negotiating mandate, on the issue of existing stocks as well as on other issues. Participation as broad as possible and in particular that of States possessing nuclear weapons is crucial for the treaty to be effective. In our view, the Group of Governmental Experts should identify all relevant issues to be addressed in the treaty, pinpoint the difficulties and focus on finding common ground. It might be worthwhile to consider establishing a committee of scientific experts to work on more technical issues and advise the negotiators before or during negotiations. We see ourselves as a potential bridge-builder and are willing to take responsibility both in the preparations and in actual negotiations.

Nigeria

[Original: English]
[14 May 2013]

During the sixty-seventh session of the General Assembly, Nigeria joined 165 countries that voted in favour of the resolution on a treaty banning the production of fissile material for nuclear weapons or other nuclear devices.

Nigeria wishes to reiterate that while such a treaty could be an important qualitative disarmament measure, it should also be stressed that the potential of this resolution might not be fully realized if the measures being considered are limited only to the future production of fissile materials while neglecting the challenges of existing stocks.

Nigeria believes that the question and challenges of existing stocks of fissile materials should be included in the task assigned to the proposed Group of Governmental Experts with a membership of twenty-five States chosen on the basis of equitable geographical representation, as recommended in the resolution. In addition, we wish to underscore that the Group's mandate could also include but not be limited to the following: all fusion and fission devices, nuclear testing, enrichment, reprocessing, separation, purification and other related issues. We are of the strong view that a holistic approach to the consideration of the preceding issues will further enrich the proposals and recommendations to be submitted to the Secretary-General, for onward transmission to the Conference on Disarmament.

Norway

[Original: English]
[14 May 2013]

Norway welcomes the decision by the First Committee to establish a working group on a treaty banning the production of fissile materials for nuclear weapons or other nuclear devices. Such a treaty could be an important contribution to promoting disarmament and preventing proliferation. It is the view of Norway that the full potential of such a treaty would be reached only if it were also to include measures regarding existing stocks of fissile materials. They would further the value of the treaty as a measure to support the goal of eliminating nuclear weapons. We are looking forward to studying the report of the working group and its recommendations to the General Assembly at its sixty-eighth session.

Pakistan

[Original: English]
[1 May 2013]

While global disarmament and arms control efforts are supposed to promote collective security, States join such processes only if the instrument or treaty under negotiation does not imperil their fundamental security interests. This has been recognized in paragraph 29 of the Final Document of the Tenth Special Session of the General Assembly.

Pakistan did not introduce nuclear weapons in South Asia. It was compelled to respond to the development of nuclear weapons by our neighbour, leading to its nuclear tests in 1974 and subsequently in 1998.

Since the idea of a treaty banning the production of fissile material for nuclear weapons or other nuclear explosive devices was introduced at the Conference on Disarmament in 1995, Pakistan has insisted that such a treaty should not only ban future production of fissile material but also address the serious asymmetry in fissile material stockpiles, especially in South Asia.

Pakistan's concerns regarding asymmetry in stocks have been further accentuated as a result of the discriminatory policies relating to selective "civilian nuclear cooperation", guided by strategic and commercial interests of some States, which has enhanced the production of fissile material for military purposes by our neighbour. In such circumstances, Pakistan has been compelled to oppose negotiations for a treaty on fissile material that would halt only future production, as that would permanently freeze its disadvantage and undermine its deterrent capability. In such a situation the so-called "constructive ambiguity" of the Shannon mandate is not sufficient to address our security concerns.

Moreover, we need to recognize that a fissile material treaty was conceived as a step towards nuclear disarmament and not merely non-proliferation. The treaty, as currently conceived for negotiations in the Conference, would have no impact on States that now have superfluous fissile material for weapons purposes. Unless such a treaty unambiguously covers the question of reducing existing stockpiles of fissile material, it would make no contribution to nuclear disarmament.

The deadlock in the Conference is not a recent development. No negotiations have taken place in the Conference since 1996 owing to differences over varying priorities. Indeed, it is due to considerations of national security of some States that no progress has been made in the Conference in over three decades to even begin negotiations on issues such as nuclear disarmament, negative security assurances and the prevention of an arms race in outer space. Pakistan's opposition to the commencement of negotiations on a fissile material treaty in the Conference on grounds of national security is, therefore, neither exceptional nor unprecedented.

Unless a level playing field is provided in South Asia, Pakistan would not be able to join negotiations on the treaty. Pakistan believes the establishment of the Group of Government Experts under the mandate of the General Assembly undermines the role of the Conference. The mandate of the proposed Group of Governmental Experts, to discuss substantive aspects of a treaty banning the production of fissile material for nuclear weapons or other nuclear explosive devices, could easily be fulfilled in the Conference. A balanced consideration of all issues on the Conference's agenda, under the well-established principle of equal and undiminished security of all States, can break the deadlock.

Peru

[Original: Spanish]
[29 April 2013]

Resolution 67/53 has revitalized the question of banning the production of fissile material to further nuclear disarmament and non-proliferation, which are laudable aims that should be supported. It is therefore crucial to discuss and clearly define the provisions of the treaty and to analyse the role of the international safeguards system, whose purpose appears to encompass implicitly the treaty's objectives.

In that connection, it is important to discuss the issues below, which should be considered in the proposal.

Nuclear material used in nuclear weapons must have basic characteristics that allow it to be used, such as the type of material, its enrichment grade and the quantity. In that connection, it is essential to define the fissile material that the treaty will cover, bearing in mind that plutonium-239 and uranium-233 are produced in a reactor and, in order to be useful, require chemical separation from fission products, while uranium-235 originating from natural uranium must be highly enriched, possibly higher than 90 per cent. Significant quantities specified by IAEA for controls are 8 kg of plutonium-239 and 25 kg of highly enriched uranium-235. These factors will therefore need to be considered.

In using nuclear material to generate electricity, plutonium-239 is created by irradiating uranium-238, which absorbs a neutron and decays to plutonium-239, and uranium-233 is formed when thorium-232 is irradiated in a reactor. Uranium-235 occurs naturally but requires enrichment. In order for them to be used, chemical separation, enrichment and reprocessing processes must be carried out, as appropriate. Merely creating this fissile material will not mean that it has been "produced", since special technological processes are required to achieve weapons-grade nuclear material. It will therefore be necessary to clearly define the scope of the word "production", as too strict a definition could unduly affect the peaceful use of nuclear material or even its use in naval reactors that require highly enriched uranium.

There are currently reserves of fissile material that could be used in various ways, including in nuclear weapons. These stocks, with a few exceptions, are not covered by safeguards. In that connection, it should be determined whether this fissile material, already produced in the past, should be covered by the treaty.

Compliance with the treaty would be subject to an additional international verification regime that would possibly be in addition to the safeguards regime. How this regime will be applied must be discussed, taking into consideration that obligations exist under the IAEA safeguards agreements, which seek, in every sense, to restrict the production of fissile material for non-peaceful purposes.

To achieve a treaty banning the production of fissile material for nuclear weapons that will contribute to nuclear disarmament and non-proliferation, it will be necessary to discuss and analyse issues related to the definition of fissile material and the word "production", whether fissile material stocks should be included and the relationship with the international IAEA safeguards regime.

Poland

[Original: English]
[10 May 2013]

Poland attaches priority to the immediate commencement of the negotiation in the Conference on Disarmament of a treaty banning the production of fissile material for nuclear weapons or other nuclear explosive devices, on the basis of document CD/1299 of 24 March 1995, which has been subsequently referred to in decision CD/1864.

We are convinced that the treaty, by banning the production of fissile material for nuclear weapons or other nuclear explosive devices, would contribute significantly to nuclear disarmament efforts under article VI of the Treaty on the Non-Proliferation of Nuclear Weapons. The treaty banning the production of fissile material for nuclear weapons or other nuclear explosive devices constitutes the next multilateral instrument to be negotiated in the nuclear disarmament field as a complement to the Non-Proliferation Treaty and the Comprehensive Nuclear-Test-Ban Treaty. The international community's support for the immediate commencement in the Conference of negotiations on a fissile material treaty in the Conference on Disarmament has been expressed on many occasions.

Poland has always actively supported the immediate commencement of treaty negotiations within the framework of the Conference, on the basis of decision CD/1864. Moreover we have also supported other solutions leading to the start of talks. Poland also took an active part in the consultations on the practical and technical aspects of the feasibility of fissile material treaty verification, which were carried out within the framework of the Non-proliferation and Disarmament Initiative under the German leadership in Vienna.

Qatar

[Original: Arabic, English]
[11 March 2013]

The State of Qatar supports the signing of a convention on the banning of the production of fissile material for nuclear weapons and other nuclear explosive devices, provided that the convention is multilateral, is not discriminatory, includes the current inventory and can be effectively verifiable at the international level. The State of Qatar believes that this convention will contribute to a large extent to the prevention of nuclear proliferation and the promotion of nuclear disarmament.

Republic of Korea

[Original: English]
[28 June 2013]

The international community should commence negotiations on a treaty banning the production of fissile material for nuclear weapons or other nuclear explosive devices at the earliest possible date, with a view to adopting a multilateral, non-discriminatory and internationally and effectively verifiable treaty.

The Republic of Korea supports the establishment of the Group of Governmental Experts in accordance with resolution 67/53, and hopes that the Group will play a valuable role in facilitating discussions and negotiations.

The Republic of Korea supports a ban on the production of fissile material for nuclear weapons or other nuclear explosive devices. Regarding the definition of fissile material that is to be banned, it would be a good approach to include therein unirradiated direct use material and other special fissionable material to be specified by the international community at a future stage.

The treaty should not hinder the production, use and development of fissile material for peaceful uses, since States have the right to produce, use, store and process fissile material for peaceful uses under the international verification system.

With regard to existing stockpiles, the Republic of Korea believes that it would be best to first commence the negotiations on a treaty, and discuss the issue of existing stockpiles at a later stage during the negotiations.

The treaty should include appropriate verification mechanisms, and the Republic of Korea supports the approach of utilizing IAEA as an inspector. In discussing concrete methodology for inspections, such factors as effectiveness and cost should be taken into account.

With regard to ratification requirements, taking into account the lessons learned from the Comprehensive Nuclear-Test-Ban Treaty, the Republic of Korea suggests that the treaty require ratification by a certain number of countries, including the five nuclear-weapon States, based on a practical approach.

Pending the adoption and implementation of a fissile material treaty, all States with nuclear weapon capabilities should voluntarily declare moratoriums on the production of fissile material for weapons purposes without further delay as an interim measure, if they have not yet done so.

Serbia

[Original: English]
[24 May 2013]

In accordance with the Treaty on the Non-Proliferation of Nuclear Weapons and its status as a non-nuclear-weapon State, Serbia does not possess nuclear weapons or equipment for the production of nuclear fission materials or nuclear explosive devices and does not plan to develop or procure equipment or materials for their production.

Serbia considers that the treaty banning the production of fissile material for nuclear weapons or other nuclear explosive devices will complement and reinforce the existing legal framework established for the purpose of disarmament and prevention of nuclear weapons proliferation. The treaty should cover both the production and the existing stockpiles of fissile material, in view of the fact that the present stocks of uranium and plutonium intended for military use are very important. The essential elements of the treaty are as follows:

- Moratorium on the production of fissile material for nuclear weapons and other nuclear explosive devices
- Establishment of a verification system, including monitoring fissile material production facilities and keeping records of those materials
- The verification system must be in place in both the nuclear-weapon States and the non-nuclear-weapon States. The latter States are already implementing this verification procedure through the Non-Proliferation Treaty by way of monitoring and control activities carried out by IAEA
- Closing as well as decommissioning military fissile material production facilities or their conversion into facilities for the production of fissile material for peaceful applications
- Providing guarantees that the fissile material produced for peaceful uses will not be used for military purposes, that is, for the production of nuclear weapons or other nuclear explosives

Particular emphasis should be placed on openness and transparency in the implementation of such a treaty to build global confidence among the States parties regarding the fissile material production issue.

South Africa

[Original: English]
[16 May 2013]

South Africa has long promoted a world without nuclear weapons. As part of the systematic and progressive approach towards achieving nuclear disarmament, South Africa is fully supportive of negotiations on a treaty that would ban the production of fissile material for nuclear weapons and other nuclear explosive devices that fulfils both nuclear non-proliferation and nuclear disarmament objectives.

While a fissile material treaty would reinforce the objectives of the Treaty on the Non-Proliferation of Nuclear Weapons and complement the Comprehensive Nuclear-Test-Ban Treaty, it would also constitute an essential element of a comprehensive framework of mutually reinforcing instruments aimed at achieving and maintaining a world without nuclear weapons.

Over the years many issues have complicated agreement on the commencement of negotiations on a fissile material treaty. Those issues have included the mandates of subsidiary bodies in the Conference on Disarmament, the vexed "linkage" question, the scope of the treaty and especially whether it should include past production and fissile material stocks, questions about its verification, as well as concerns regarding the utility of a cut-off treaty as a genuine nuclear disarmament step that will not merely freeze the status quo.

As the first country to have developed and then completely eliminated its nuclear weapons, South Africa is fully aware of the complexities associated with a future treaty. Contrary to arguments about the limitations of such a future, the experience of South Africa has shown that despite significant technical complexities that will need to be acknowledged and addressed, all of them can be overcome with the necessary political will. While acknowledging the difficulties associated with the past production of fissile material, South Africa strongly believes that the question of stocks needs to be addressed in a future treaty for it to be a credible disarmament instrument. A fissile material treaty that fulfils disarmament objectives will necessarily also have to give effect to the principles of transparency, irreversibility and verification agreed to during the review conferences of the parties to the Treaty on the Non-Proliferation of Nuclear Weapons in 2000 and 2010.

For such a treaty to be fully effective, we believe that it should be the product of multilateral disarmament negotiations. The non-nuclear-weapon States under the Non-Proliferation Treaty already have a verifiable obligation not to produce fissile material for nuclear weapons purposes. The conclusion of an agreement without the participation of those States that do not have existing obligations would therefore not be meaningful. Although South Africa remains ready to commence negotiations immediately on such a treaty, it does not subscribe to the view that this is the only issue "ripe" for negotiation, or that the commencement of such negotiations should become a prerequisite for further progress on nuclear disarmament. As a meaningful disarmament step, a fissile material treaty cannot be separated from the overall objective, which is and should remain a comprehensive framework of mutually reinforcing instruments aimed at the establishment and maintenance of a world without nuclear weapons.

In 2002, South Africa submitted a working paper to the Conference on a fissile material treaty (CD/1671). Building upon that paper, South Africa's reflections on the possible scope and requirements of a fissile material treaty have been submitted as a contribution to the work of the Group of Governmental Experts that will be convened in 2014 (full text available from the website of the Conference; see para. 4).

Sweden

[Original: English]
[16 May 2013]

Sweden regards a treaty banning the production of fissile material for nuclear weapons or other nuclear explosive devices as a key disarmament and non-proliferation element, integral to any framework of mutually reinforcing legal instruments for achieving and maintaining a world without nuclear weapons. Looking forward to the establishment of the Group of Governmental Experts, Sweden would be ready to take an active part in the deliberations of the Group. With significant competence and experience relating to, inter alia, technical verification work, Sweden considers itself well equipped to contribute to efforts towards preparing the ground for the treaty, both technologically and otherwise. Sweden has continuously addressed issues related to the treaty and would like to provide preliminary and non-exhaustive comments on some central treaty-related issues as "food for thought" for future discussions. It will be important to define and agree on how its expected utility can best be maximized by a future treaty. While discussions in the Group are likely to be broad, it is possible to identify several important technical and organizational issues that the Group could address, and in so doing create a more solid basis for future substantive work. This would include a careful assessment of the feasibility of necessary verification measures. The work of the Group could make an important contribution to further clarification of technical and other implications of a variety of scopes and definitions of fissile materials in a treaty, including the issue of addressing existing stocks of fissile material and how such stocks could be defined and classified.

Relevant technical issues of this type include, for example:

- Discussion of possible definitions of military and civilian stocks. The choice of definitions will impact the effectiveness and/or intrusiveness of the treaty verification regime. The verification of fissile material flows between facilities using different stock definitions could also be investigated.
- The term "fissile material" is not formally defined or used in any verification regime. Possible definitions of fissile material and their implications for treaty verification and scope are among crucial issues that the Group could explore.
- What is meant by "production" of fissile material?
- The verification regime would most likely include many of the verification techniques already part of the IAEA nuclear safeguards toolbox, tailored to fit a fissile material treaty. There are also, however, verification techniques used in other applications or under development that could be further investigated with respect to future treaty verification. This is particularly true for verification of clandestine production of fissile material. Ideally, the outcome of the work of the Group should result in a set of recommendations for the scope, definitions and verification of a future treaty.

Switzerland

[Original: English]
[6 June 2013]

A treaty on fissile material for nuclear weapons is long overdue and represents a priority for Switzerland. Reaching an understanding on modalities should not be made a precondition for beginning negotiations. A treaty should strengthen and complement the existing nuclear non-proliferation and disarmament regime. It should halt both vertical and horizontal proliferation and contribute to nuclear disarmament. Consequently, a treaty should, on the one hand, prohibit the future production of fissile material for nuclear weapons or other nuclear explosive devices, including providing for the decommissioning and dismantlement of production facilities or for their reconfiguration for peaceful purposes only.

A treaty should, on the other hand, address past production of fissile material. If only future production is covered (in a mere "cut-off treaty"), the disarmament effects will be limited. Such an approach could generate incentives for a State to produce as much material as possible before ratifying the treaty or, worse, even create incentives to delay the commencement of negotiations or the entry into force of such a treaty. By including coverage of existing stock in the treaty, nuclear disarmament will be advanced significantly, not the least by ensuring that existing fissile material excess to military requirements will never return to nuclear weapons stocks. A treaty should also ensure that stocks of highly enriched uranium for naval propulsion will not and cannot be used in nuclear weapons and other explosive devices.

Syrian Arab Republic

[Original: Arabic]
[14 May 2013]

With regard to a treaty banning the production of fissile material for nuclear weapons or other nuclear explosive devices, the Syrian Arab Republic wishes to affirm the following:

1. The Conference on Disarmament is the single multilateral disarmament negotiating forum of the international community.

2. The treaty must be non-discriminatory, international and multilateral, and must apply on an equal basis to all States that produce, store, possess or transfer such material. In other words, the treaty must be a non-discriminatory, international instrument.

3. A non-discriminatory and transparent international mechanism must be established to monitor the production and storage of such material.

4. The treaty must contain explicit provisions concerning a mechanism for eliminating the fissile material stockpiles of nuclear-weapon States or those of other States, regardless of how they obtained such material.

Ukraine

[Original: English]
[3 June 2013]

Ukraine strongly supports full and effective implementation of legal and institutional multilateral mechanisms to prevent the proliferation of nuclear weapons and related materials, equipment and technologies.

Ukraine continues to support the early commencement of international negotiations on a treaty banning the production of fissile material for nuclear weapons or other nuclear explosive devices in the Conference on Disarmament. Until then, we call upon all States concerned to declare and uphold a moratorium on the production of such material.

Ukraine considers that a fissile material treaty should reflect a balance between nuclear disarmament and non-proliferation dimensions and that the issue of existing stocks of fissile materials should be addressed during the negotiation process, and should not be used to block any disarmament negotiating forum.

A draft treaty should include the following principal provisions:

- Prohibition of the production of highly enriched uranium (contains not less than 20 per cent of the uranium 235 isotope) and plutonium (the plutonium 238 isotope content is less than 80 per cent)
- An effective regime to verify compliance with the treaty, including submission of initial declarations, regular inspections of declared facilities, as well as inspections on demand to detect possible prohibited activities at undeclared objects
- Initial declaration of all objects of enrichment and reprocessing of uranium and plutonium, regardless of their current status (active, closed, dismantled or converted facilities)

- Ensure cost-effectiveness through the exclusion of all completely dismantled objects from the inspection regime
- Inspections on demand for detection of possible prohibited activities at undeclared objects should be carried out in accordance with the controlled access procedure to prevent leakage of sensitive information relating to nuclear non-proliferation or national security interests
- A future inspection regime should not create any additional obligations in this field for non-nuclear States that are parties to the IAEA comprehensive safeguards agreements.

United Kingdom of Great Britain and Northern Ireland

[Original: English]
[16 May 2013]

The United Kingdom of Great Britain and Northern Ireland supported the resolution to create the fissile material cut-off treaty Group of Government Experts at a meeting of the First Committee of the General Assembly in 2012. We look forward to a constructive Group that will help to set some of the parameters for a treaty, including the development of a solid technical framework to support key elements of the future treaty.

We believe that the Group will complement existing efforts to find a positive way forward in the Conference on Disarmament. The United Kingdom looks forward to participating in the Group and believes that it should consider the following:

The precise scope of the treaty, including:

- The definition of "fissile material" to be used by the treaty
- How "production" of fissile materials could best be defined
- Whether or not existing stocks of fissile material should be covered in some way

The most appropriate verification arrangements, including:

- The general approach for verification of the treaty
- Which body or bodies would carry out the verification
- How verification arrangements are treated in the treaty itself

The provisions governing the operation of the treaty itself:

- The treaty will need to outline its entry-into-force requirements, its duration, and any arrangements for review, amendment and withdrawal.

Our note to the United Nations sets out the views of the United Kingdom on each of the above-mentioned issues in fuller detail (available from the website of the Conference; see para. 4).

United States of America

[Original: English]
[10 May 2013]

The United States of America believes that achieving a treaty banning the production of fissile material for nuclear weapons or other nuclear explosive devices would be an important milestone for nuclear non-proliferation as well as the next logical step in multilateral nuclear disarmament. The fundamental obligation of the treaty would be to ban the production of fissile material for use in nuclear weapons or other nuclear explosive devices. "Fissile material" should be defined to correspond to what IAEA calls "direct-use material". "Production" of fissile material should be defined as occurring through isotopic separation—enrichment—or through chemical separation of irradiated nuclear material—reprocessing. Therefore "production facilities" would correspond to enrichment and reprocessing facilities.

A treaty verification regime would need to be negotiated as part of the treaty. All production facilities would need to be declared under the treaty, and all newly produced fissile material declared, accounted for and monitored to verify that the materials are not being diverted for use in weapons. Verification would also need to include detection of undeclared production facilities. For that purpose, measures beyond routine monitoring would be needed. A "managed access" protocol that protects sensitive information would need to be developed for that purpose.

The United States believes that IAEA is best suited and should be responsible for carrying out treaty monitoring and inspections. However, there would remain a need for higher-level decision-making on important treaty issues, such as compliance questions and the reviewing of treaty implementation. The Agency may or may not be the appropriate organization for those functions, and an organization consisting of parties to the treaty could be established.

Legitimate points for negotiation include the scope of the treaty, explicitly recognized in the Shannon mandate. The United States believes that the treaty should not place legal obligations on existing fissile material. Inclusion of existing stocks would make the treaty much more difficult to negotiate successfully. In addition, separate from a fissile material treaty, much progress has already been made, and still is being made on measures to eliminate existing stocks. Further progress could be hampered by linkage to such a treaty.

Uruguay

[Original: Spanish]
[1 May 2013]

In response to the Secretary-General's request, made in accordance with General Assembly resolution 67/53, entitled "Treaty banning the production of fissile material for nuclear weapons or other nuclear explosive devices", it should be recalled that Uruguay supported that resolution, as it is a step, however small, towards initiating negotiations to agree on that instrument, which would undoubtedly be a key contribution to the cause of nuclear disarmament and non-proliferation. The instrument could also strengthen the Treaty on the Non-Proliferation of Nuclear Weapons.

Generally speaking, a treaty of this nature would make it possible to coordinate or integrate on a technical foundation various viable, or at least feasible, political initiatives to secure, consolidate and reduce reserves of highly enriched uranium and plutonium. It is well known that the control of fissile material is crucial, not only for nuclear disarmament, but also to stop the proliferation of nuclear weapons and ensure that terrorists cannot acquire them. Therefore, the fact that some States not party to the Treaty on the Non-Proliferation of Nuclear Weapons have continued to produce plutonium and highly enriched uranium has been and continues to be a cause for concern.

It is clear that the ban on the production of fissile material should apply only to material used directly in weapons or other nuclear explosive devices. It could thus be recognized that some States may separate plutonium for recycling in civilian nuclear power reactors. The IAEA safeguards would have to be applicable to all fissile material. As is known, material such as neptunium-237 and americium-241 and -243 are not included in the "special fissionable material" category of the IAEA Statute, nor are they the subject of its safeguards regime. For that reason, the definition of fissile materials must be expanded to include the aforementioned material, and any other that can sustain a fission chain reaction.

While States parties might be required to accept the necessary safeguards to ensure that they comply with the main obligations under the treaty, it should be noted that compliance with some of the obligations laid down in the treaty cannot be easily verified using the safeguards.

Another matter that should be highlighted and retained in the proposed treaty is the possible adoption of protocols, which could facilitate contemplation of new details concerning the implementation of the proposed treaty or allow subgroups of countries to undertake additional commitments without having to amend the text itself. Consequently, depending on the circumstances, protocols could be applicable to all States parties or to a particular group.

Negotiating the treaty has not been and will not be easy. Its achievement would not only limit the materials available for the manufacture of nuclear weapons, but would also significantly strengthen measures governing disarmament, non-proliferation and arms control. Moreover, it would be a decisive step towards improving the climate of trust in a field of international security fraught with rising tensions.

III. Reply received from the European Union

[Original: English]
[31 May 2013]

For the European Union, the immediate commencement and early conclusion of the negotiation in the Conference on Disarmament of a treaty banning the production of fissile material for nuclear weapons or other nuclear explosive devices, on the basis of document CD/1299 and the mandate contained therein, remains a clear priority. Such a treaty constitutes an urgent necessity in the nuclear disarmament field as a complement to the Treaty on the Non-Proliferation of Nuclear Weapons and the Comprehensive Nuclear-Test-Ban Treaty.

National security concerns, while legitimate, can and should be addressed as part of the negotiation process rather than as a prerequisite. Confidence-building measures can be taken immediately, without the need to wait for the commencement of formal negotiations. In that regard, pending negotiations and the entry into force of the treaty, the European Union calls on all States concerned to declare and uphold an immediate moratorium on the production of fissile material for nuclear weapons or other nuclear explosive devices.

All European Union member States supported General Assembly resolution 67/53. The mechanism established by the resolution represents a useful contribution to helping the Conference without undermining its authority and primary role in multilateral disarmament negotiations.

The technical expert meetings on the treaty organized by two European Union member States, Germany and the Netherlands, in May and August 2012, respectively, in support of the early commencement of negotiations within the Conference were useful; they enhanced our knowledge and understanding of technical issues.

The European Union reaffirms its strong commitment to the Conference as the single multilateral disarmament negotiating forum of the international community. The Conference, in accordance with its mandate, has the crucial role of negotiating multilateral treaties. Its ongoing stalemate remains deeply troubling. Adopting and implementing a programme of work will, inter alia, enable negotiations on a treaty.